T0202902

Communications
in Computer and Information Science   1883

Rationale

The CCIS series is devoted to the publication of proceedings of computer science conferences. Its aim is to efficiently disseminate original research results in informatics in printed and electronic form. While the focus is on publication of peer-reviewed full papers presenting mature work, inclusion of reviewed short papers reporting on work in progress is welcome, too. Besides globally relevant meetings with internationally representative program committees guaranteeing a strict peer-reviewing and paper selection process, conferences run by societies or of high regional or national relevance are also considered for publication.

Topics

The topical scope of CCIS spans the entire spectrum of informatics ranging from foundational topics in the theory of computing to information and communications science and technology and a broad variety of interdisciplinary application fields.

Information for Volume Editors and Authors

Publication in CCIS is free of charge. No royalties are paid, however, we offer registered conference participants temporary free access to the online version of the conference proceedings on SpringerLink (http://link.springer.com) by means of an http referrer from the conference website and/or a number of complimentary printed copies, as specified in the official acceptance email of the event.

CCIS proceedings can be published in time for distribution at conferences or as post-proceedings, and delivered in the form of printed books and/or electronically as USBs and/or e-content licenses for accessing proceedings at SpringerLink. Furthermore, CCIS proceedings are included in the CCIS electronic book series hosted in the SpringerLink digital library at http://link.springer.com/bookseries/7899. Conferences publishing in CCIS are allowed to use Online Conference Service (OCS) for managing the whole proceedings lifecycle (from submission and reviewing to preparing for publication) free of charge.

Publication process

The language of publication is exclusively English. Authors publishing in CCIS have to sign the Springer CCIS copyright transfer form, however, they are free to use their material published in CCIS for substantially changed, more elaborate subsequent publications elsewhere. For the preparation of the camera-ready papers/files, authors have to strictly adhere to the Springer CCIS Authors' Instructions and are strongly encouraged to use the CCIS LaTeX style files or templates.

Abstracting/Indexing

CCIS is abstracted/indexed in DBLP, Google Scholar, EI-Compendex, Mathematical Reviews, SCImago, Scopus. CCIS volumes are also submitted for the inclusion in ISI Proceedings.

How to start

To start the evaluation of your proposal for inclusion in the CCIS series, please send an e-mail to ccis@springer.com.

Mehdi Ghatee · S. Mehdi Hashemi
Editors

Artificial Intelligence and Smart Vehicles

First International Conference, ICAISV 2023
Tehran, Iran, May 24–25, 2023
Revised Selected Papers

 Springer

Editors
Mehdi Ghatee ⓘ
Amirkabir University of Technology
Tehran, Iran

S. Mehdi Hashemi
Amirkabir University of Technology
Tehran, Iran

ISSN 1865-0929   ISSN 1865-0937 (electronic)
Communications in Computer and Information Science
ISBN 978-3-031-43762-5   ISBN 978-3-031-43763-2 (eBook)
https://doi.org/10.1007/978-3-031-43763-2

This Springer imprint is published by the registered company Springer Nature Switzerland AG
The registered company address is: Gewerbestrasse 11, 6330 Cham, Switzerland

Paper in this product is recyclable.

Preface

On May 24–25, 2023, the Amirkabir University of Technology organized the first International Conference on Artificial Intelligence and Smart Vehicles (ICAISV 2023) in collaboration with different scholars from worldwide universities and research institutes. A scientific committee consisting of 63 scientific experts from Iran, USA, UK, Canada, Australia, Southeast Asia, and Europe accompanied us to hold this scientific event. We organized a forum for researchers and engineers to collaborate on artificial intelligence technologies for smart vehicles and intelligent transportation systems. We focused on machine learning, data mining, machine vision, image processing, signal analysis, decision support systems, expert systems, and their applications in smart vehicles. Knowledge extraction from data captured by vehicles, cellphones, smartwatches, and sensor networks of V2V, V2I, and V2X connections was one of our attention points. The majority of articles received were in the fields of autonomous vehicles, image processing, and connected vehicles. To review the 93 articles received by the conference office, 178 reviewers with multi-disciplinary specialties of computer science and engineering, electrical engineering, mechanical engineering, industrial engineering, civil engineering, mathematics, and management were selected and 602 invitation letters were sent. Based on the received reviewing reports, 35 articles were accepted for presentation at the conference. These papers were presented in 13 hybrid sessions to more than 270 conference participants. The 14 top papers were accepted to be published in this volume of Springer's Communications in Computer and Information Science (CCIS) book series. Also, the following key speakers presented their latest achievements in the conference sessions:

- Mahdi Rezaei
  Associate Professor with Institute for Transport Studies, University of Leeds, UK
- Nima Mohajerin
  Senior Research Lead at Nuro AI, US
- Mohammad Pirani
  Research Assistant Professor with the Department of Mechanical and Mechatronics Engineering, University of Waterloo, Canada
- Hamze Zakeri
  Head of the Center for the Development of Education, Studies, and Innovation, Iran Road Maintenance & Transportation Organization, Iran
- Mohammad Mahdi Bejani
  Principal Researcher with the Department of Computer Science, Faculty of Mathematics and Computer Science, Amirkabir University of Technology (Tehran Polytechnic), Iran

In addition to many guests from universities and industries, Ahmad Vahidi (the Minister of the Interior of Iran), Shahriar Afandizadeh (Deputy Minister of Roads and Urban Development for Transportation Affairs, Iran), Dariush Amani (Deputy Minister of Roads and Urban Development and Head of Iran Road Maintenance & Transportation

Organization, Iran), and Abbas Aliabadi (Minister of Industry, Mining and Trade, Iran) supported the conference and presented special talks at the conference. We sincerely appreciate all the support, participants, speakers, and guests of the conference. We also warmly acknowledge Springer's team for their support and publication of the top papers.

May 2023   Mehdi Ghatee
                                                                S. Mehdi Hashemi

Organization

Program Committee Chairs

Mehdi Ghatee   Amirkabir University of Technology, Iran
S. Mehdi Hashemi   Amirkabir University of Technology, Iran

Organizing Committee

Seyed Hassan Ghodsypour	Chancellor of Amirkabir University of Technology, Amirkabir University of Technology, Iran
Mostafa Keshavarz Moraveji	Director General of the Chancellor's Office, Amirkabir University of Technology, Iran
Mohammad Javad Ameri Shahrabi	Vice Chancellor for Research and Technology, Amirkabir University of Technology, Iran
Dariush Kiani	Dean of Department of Mathematics and Computer Science, Amirkabir University of Technology, Iran
Mehdi Ghatee (Conference Chair)	Amirkabir University of Technology, Iran

Scientific Committee

Farshid Abdollahi	Shiraz University, Iran
Monireh Abdoos	Shahid Beheshti University, Iran
Shadi Abpeikar	University of New South Wales Canberra, Australia
Lounis Adouane	CNRS, Université de Technologie de Compiègne, France
Samira Ahangari	Morgan State University, USA
Abbas Ahmadi	Amirkabir University of Technology, Iran
Mohammad Akbari	Amirkabir University of Technology, Iran
Imad L. Al-Qadi	University of Illinois at Urbana-Champaign, USA
Vijay Anant Athavale	Walchand Institute of Technology, India
Reza Asadi	Uber AI, USA
HamidReza Attaeian	Sharif University of Technology, Iran
Mohammad Mahdi Bejani	Amirkabir University of Technology, Iran

Cheng Siong Chin	Newcastle University in Singapore, Singapore and Chongqing University, China
Paolo Dabove	Politecnico di Torino, Italy
Masoud Dahmardeh	Iran University of Science and Technology, Iran
HamidReza Eftekhari	Malayer University, Iran
Foad Ghaderi	Tarbiat Modares University, Iran
Amirhossein Ghasemi	University of North Carolina Charlotte, USA
Mehdi Ghatee	Amirkabir University of Technology, Iran
S. Ali Ghorashi	University of East London, UK
Amir Golroo	Amirkabir University of Technology, Iran
Luis Carlos Gonzalez-Gurrola	Universidad Autónoma de Chihuahua, Mexico
Brij Bhooshan Gupta	Asia University, Taiwan (R.O.C.)
Joko Hariyono	Sebelas Maret University, Indonesia
S. Mehdi Hashemi	Amirkabir University of Technology, Iran
Yeganeh Hayeri	Stevens Institute of Technology, USA
Alireza Jolfaei	Flinders University, Australia
Behrooz Karimi	Amirkabir University of Technology, Iran
Ali Kashif Bashir	Manchester Metropolitan University, UK
Anupam Kumar	Delft University of Technology, The Netherlands
Zhen Leng	Hong Kong Polytechnic University, China
Fernando Lezama	GECAD - Research Group on Intelligent Engineering and Computing for Polytechnic Institute of Porto, Portugal
Farhad Maleki	University of Calgary, Canada
Hormoz Marzbani	RMIT University, Australia
Fereidoon Moghadas Nejad	Amirkabir University of Technology, Iran
Nima Mohajerin	Microsoft, Canada
Ali Movaghar	Sharif University of Technology, Iran
Omid Naghshineh	Amirkabir University of Technology, Iran
Ali Nahvi	K.N. Toosi University of Technology, Iran
Abdolreza Ohadi Hamedani	Amirkabir University of Technology, Iran
Mohammad Oskuoee	Niroo Research Institute, Iran
Md. Jalil Piran	Sejong University, South Korea
Mohammad Pirani	University of Waterloo, Canada
Alireza Rahai	Amirkabir University of Technology, Iran
Zahed Rahmati	Amirkabir University of Technology, Iran
Sérgio Ramos	Polytechnic Institute of Porto, Portugal
Mehdi Rasti	Amirkabir University of Technology, Iran
Mahdi Rezaei	University of Leeds, UK
Mansoor Rezghi	Tarbiat Modares University, Iran
Majid Rostami-Shahrbabaki	Technical University of Munich, Germany
Ramin Saedi	Amazon Inc., USA

Mahmoud Saffarzadeh	Tarbiat Modares University, Iran
Ahmad Sepehri	Amirkabir University of Technology, Iran & CAPIS, Belgium
Majid Shalchian	Amirkabir University of Technology, Iran
Saeed Sharifian	Amirkabir University of Technology, Iran
Mohammad Hassan Shirali-Shahreza	Amirkabir University of Technology, Iran
Omar Smadi	Iowa State University, USA
João Soares	GECAD - Research Group on Intelligent Engineering and Computing for Polytechnic Institute of Porto, Portugal
Sadegh Vaez-Zadeh	University of Tehran, Iran
Johan Wahlström	University of Exeter, UK
Michael P. Wistuba	Technische Universität Braunschweig, Germany
Halim Yanikomeroglu	Carleton University, Canada
Muhammad Aizzat Bin Zakaria	Universiti Malaysia Pahang, Malaysia

Reviewers

Elham Abbasi
Mostafa Abbaszadeh
Arash Abdi
Maryam Abdolali
Farshid Abdollahi
Monireh Abdoos
Shadi Abpeikar
Hojjat Adibi
Lounis Adouane
Ahmad Afshar
Samira Ahangari
Abbas Ahmadi
Ruhallah Ahmadian
Amir Ahmadijavid
Mohammad Akbari
Imad L. Al-Qadi
Fadi Al-Turjman
Pooyan Alavi
Mohammad Ali
Mohammad Javad Ameri Shahrabi
Maryam Amirmazlaghani
Reza Asadi
Vijay Anant Athavale
Hamidreza Attaeian

Amirreza Babaahmadi
Sachin Babar
Alireza Bagheri
Saeed Bagheri
Ali Kashif Bashir
Mohammad Mahdi Bejani
Milad Besharatifar
Mohamadmahdi Bideh
Muhammad Aizzat Bin Zakaria
Neda Binesh
Shahram Bohluli
Cheng Siong Chin
Paolo Dabove
Masoud Dahmardeh
Mohsen Ebadpour
Hamidreza Eftekhari
Farshad Eshghi
Hashem Ezzati
Ashkan Fakhri
Mahshid Falsafi
Ali Fanian
Vista Farahifar
Jyotsna Garikipati
Foad Ghaderi

Alireza Ghanbari
Hamid Gharagozlou
Amirhossein Ghasemi
Mehdi Ghatee
Amin Gheibi
Mehdi Gheisari
S. Ali Ghorashi
Zahra Ghorbanali
Fateme Golivand
Amir Golroo
Luis Carlos Gonzalez-Gurrola
Mateusz Goralczyk
Brij Bhooshan Gupta
Meeghat Habibian
Joko Hariyono
S. Mehdi Hashemi
Alireza Hashemigolpaigani
Hassan Hassan Ghodsyour
Yeganeh Hayeri
Marzi Heidari
Mohammadjavad Hekmatnasab
Farnaz Hooshmand
Fateme Hoseynnia
Mahdi Jafari Siavoshani
Majid Jahani
Amirhosein Jamali
Mahdi Javanmardi
Alireza Jolfaei
Ali Kamali
Farzad Karami
Behrooz Karimi
Mohammad Karimi
Manoochehr Kelarestaghi
Masoume Khodaei
Mohammad Khosravi
Dariush Kiani
Anupam Kumar
Zhen Leng
Fernando Lezama
Mehdi Majidpour
Farhad Maleki
Hormoz Marzbani
Mahmoud Mesbah
Danial Mirizadeh
Hamid Mirzahossein

Fereidoon Moghadas Nejad
Ali Mohades
Nima Mohajerin
Adel Mohamadpour
Reza Mohammadi
Saeedeh Momtazi
Mahmoud Momtazpour
Sepanta Montazeri
Morteza Moradi
Parham Moradi
Sepehr Moradi
Hadi Mosadegh
Saeedeh Mosaferchi
Behzad Moshiri
Ahmad Motamedi
Ali Movaghar
Omid Naghshineh
Ali Nahvi
Babak Najjar Araabi
Samad Najjar-Ghabel
Mahyar Naraghi
Mohammadjavad Nazari
Ehsan Nazerfard
Ahmad Nickabadi
Salman Niksefat
Abdolreza Ohadi Hamedani
Mohammad Oskuoee
Ali Pashaei
Mehdi Paykanheyrati
Md. Jalil Piran
Mohammad Pirani
Arash Pourhasan
Niladri Puhan
Mehdi Rafizadeh
Alireza Rahai
Shiva Rahimipour
Amin Rahmani
Zahed Rahmati
Sérgio Ramos
Mehdi Rasti
Mahdi Rezaei
Mansoor Rezghi
Majid Rostami-Shahrbabaki
Bahram Sadeghi Bigham
Babak Sadeghian

Contents

xiv   Contents

Local and Global Contextual Features Fusion for Pedestrian Intention Prediction

Mohsen Azarmi⬥, Mahdi Rezaei$^{(\boxtimes)}$⬥, Tanveer Hussain, and Chenghao Qian

Institute for Transport Studies, University of Leeds, Leeds, UK
m.rezaei@leeds.ac.uk
https://environment.leeds.ac.uk/transport/staff/9408/dr-mahdi-rezaei

Abstract. Autonomous vehicles (AVs) are becoming an indispensable part of future transportation. However, safety challenges and lack of reliability limit their real-world deployment. Towards boosting the appearance of AVs on the roads, the interaction of AVs with pedestrians including "prediction of the pedestrian crossing intention" deserves extensive research. This is a highly challenging task as involves multiple non-linear parameters. In this direction, we extract and analyse spatio-temporal visual features of both pedestrian and traffic contexts. The pedestrian features include body pose and local context features that represent the pedestrian's behaviour. Additionally, to understand the global context, we utilise location, motion, and environmental information using scene parsing technology that represents the pedestrian's surroundings, and may affect the pedestrian's intention. Finally, these multi-modality features are intelligently fused for effective intention prediction learning. The experimental results of the proposed model on the JAAD dataset show a superior result on the combined AUC and F1-score compared to the state-of-the-art.

Keywords: Pedestrian Crossing Intention · Pose Estimation · Semantic Segmentation · Pedestrian Intent Prediction · Autonomous Vehicles · Computer Vision · Human Action Prediction

1 Introduction

Pedestrian crossing intention prediction or *Pedestrian Intention Prediction* (PIP) is deemed to be important in the context of autonomous driving systems (ADS). During the past decade, a variety of approaches have investigated similar challenging tasks and recently more studies have been conducted about pedestrian crossing behaviours [1,2] in the computer vision community.

This includes interpreting the upcoming actions of pedestrians which involve a high degree of freedom and complexity in their movements [3]. Pedestrians can select any trajectories and might show very agile behaviours or change their motion direction abruptly. Pedestrians may not follow designated crossing areas or zebra-crossing [4], and also be distracted by talking or texting over a phone or

M. Ghatee and S. M. Hashemi (Eds.): ICAISV 2023, CCIS 1883, pp. 1–13, 2023.
https://doi.org/10.1007/978-3-031-43763-2_1

with other accompanying pedestrians. Their intention for crossing could also be affected by many other factors such as traffic density [4], demographics, walking in a group or alone, road width, road structure, and many more [5]. All these factors form contextual data for PIP. However, most of the studies such as [6,7] have tried to investigate the relationship between only one or two of these factors and pedestrian crossing behaviour.

AVs consider a conservative approach in interaction with pedestrians as the most vulnerable road users (VRUs), by driving at a slow pace, avoiding complex interactions, and stopping often to avoid any road catastrophe. There is also a preference to drive in less complicated environments in terms of understanding VRUs, which limits the AVs with a low level of autonomy to participate in numerous everyday traffic scenarios [8]. However, high levels of autonomy (i.e., levels 4 and 5) demand a higher level of interaction with VRUs. A robust PIP model can provide the information needed to realise what exactly a pedestrian is about to do in a particular traffic scenario. Even for level 3 (conditional automation) vehicles, a PIP model buys more time for the vehicle and/or driver to take decisions and leads to a safer manoeuvre.

Action prediction models [9], analyse current and past video sequences of a human (temporal analysis), to estimate the human action in the forthcoming frames, or for pedestrian motion estimation. Algorithms that utilise contextual information from the scene, road infrastructure, and behavioural characteristics (e.g. pose) perform better in understanding the pedestrian's intention. Temporal analysis methods rely on the past moving trajectories of pedestrians to anticipate future action of whether a pedestrian crosses or not [10].

This study aims to consider different sources of features that could be extracted from sequential video data and pedestrian pose to develop a pedestrian crossing intention prediction model using vision-based convolutional neural networks. The model uses spatio-temporal features, pose, and contextual information extracted from the front-view camera in JAAD dataset [11].

2   Related Works

Pedestrian intention prediction (PIP) requires accurate detection [12], tracking [13], localisation [14], and moving trajectory estimation [15]. With the development of high-resolution sensors, PIP research has become more feasible and attractive.

The study by Rasouli and Toostsos [16] conduct the interaction between AVs and pedestrians in traffic scenes. They emphasise the importance of AVs' communication with other road users, and they point out the major practical and theoretical challenges of this subject. A set of other studies propose deep learning-based approaches for action recognition [3] using multi-task learning to predict the crossing intention of pedestrians [17,18]. Other studies such as [19], considered both moving trajectories and pedestrian' body pose with promising results in classifying pedestrian crossing behaviour. Although they can find an apparent relationship between body gestures and the tendency to cross, insufficient consideration of spatial features prevent them from precise intention prediction in the specialised datasets and benchmarks [20].

Another research [11] uses contextual information by extending the previous work by [5], which suggests that the crossing intention ought to be considered as a context-aware problem. They suggest considering not only the pedestrian features but also the road semantic information.

Several deep learning methods, such as 2D [21] and 3D convolutional neural networks [6,22], Long-Sort Term Memory (LSTM) [23,24], attention mechanism [25], Transformer [26], and graph neural networks [25,27] have been utilised to assess the spatio-temporal features and to correlate pedestrian features with semantic information. The outcome of more successful research works shows that using multiple sources of information, such as different sensors [28], moving trajectories [29], body gesture [30,31], semantic segmentation [32], dynamic and motion information [33,34] would lead to more accurate results. Hence, some of the recent research direction has focused on finding optimal feature fusion strategies [35] but are still in infancy.

As a new contribution, this study aims to enhance the fusion approaches by incorporating environmental features and camera motion as global features and pedestrian attributes as local features to classify and predict the crossing intention. This is evaluated against the challenging dataset of *Joint Attention in Autonomous Vehicle (JAAD)* [11].

3 The Proposed Methodology

The proposed method analyses sequential video information, the global context of the scene, and pose of the pedestrian as the input features, and as a result, predicts the final pedestrian's crossing intention or action.

We define the final action of a pedestrian as a binary classification problem consisting of two classes of "crossing" or "not crossing".

Figure 1 demonstrates the model architecture and its components, which intakes the extracted features from the video footage and generates the intention classification result as the output. The proposed architecture and the types of features are explained in subsequent sections.

3.1 Global Context

Semantic Segmentation: In order to enrich our model with global information surrounding the pedestrian, we parse the scene with a semantic segmentation algorithm which also reduces the noise of perceiving spatial information in addition to segmenting various classes of objects in the scene.

The exploited segmentator (S) interprets the input image $I_{RGB(h \times w)}$, where h, w are the height and width of the image, respectively, to produce output image $I_s(h \times w \times n) = S(I_{RGB(h \times w)})$, which contains n binary masks and each one refers to the existence of an individual class in the scene. The scene is categorised

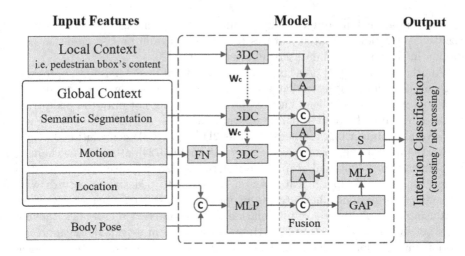

Fig. 1. The overview of the proposed model. Abbreviations: **3DC** is the 3D convolutional layer; W_c is the shared weights of the network; **MLP** is multi-layer perceptron; **FN** is Flownet2 model; **A** is self-attention module; **GAP** is global average pooling; **S** is Softmax function, and **C** refers to the feature concatenation over source dimension.

into eight classes of objects including pedestrians, roads, vehicles, constructions, vegetation, sky, buildings, and unrecognised objects, using the model proposed by [36].

Motion: We consider dense optical flow as a scene motion descriptor (F) to obtain velocity measures of the pedestrians in the dataset. However, we avoid using the classical optical flow method which compares the intensities within a given window. Instead, we utilise a deep learning-based technology, Flownet2 [37], which leverages the accuracy and accelerates the run-time performance. The network produces the output as $I_o(h \times w) = F(I_{RGB(h \times w)})$.

Location: Like other state-of-the-art models [22,25,27], we extract the location of the pedestrians directly from the JAAD ground truth; although this can be done by a detection algorithm such as [12]. JAAD's annotation provides the corresponding coordinates of the top-left and bottom-right corners of the bounding box. To avoid varying scales of input data, we adopt *min-max* scaling approach which normalises the bounding box coordinate values between 0 to 1 as follows:

$$\chi = \frac{\chi - \chi_{min}}{\chi_{max} - \chi_{min}}, \tag{1}$$

where x and y value ranges from 0 to 1920 and 0 to 1080, respectively. Due to the negligible variation of width (w_b) for bounding boxes in the dataset, we follow [26] study and remove w_b, so, the arrangement of coordinates is considered as (x, y, h_b), where the first and second elements are the centre coordinates of the bounding box, respectively, and h_b is denoted as the height of the bounding box. Thereafter, the coordinates input vector containing the location of the pedestrian is defined as $v_b \in \mathbb{R}^{B \times N \times 3}$, where B is the batch size and N is the length of the sequence of input data.

3.2 Local Context

Robust features of pedestrians including bounding box and pose information play a key role in designing a generalised model and should not be neglected. This gives the model the capability of producing accurate and precise predictions [38].

We define "Local context" as a region that constitutes the content of the pedestrian image as the shape $I_l \in \mathbb{R}^{B \times N \times h_b \times w_b \times 3}$, where 3 is the number of channels for the red, green, and blue intensities, ranging $[0, 255]$. We also normalise this vector between 0 and 1 values using Eq. 1

3.3 Body Pose

The 2D coordinates of pose key points are obtained as the output of a pose estimator algorithm as used in [31]. The pose information includes 36 values corresponding to 18 pairs of 2D coordinates, representing the body joints of the pedestrian. We have also normalised them using *min-max* scaling (Eq. 1) which results in values between 0 and 1. The vector containing the body pose is defined as $v_p \in \mathbb{R}^{B \times N \times 36}$.

3.4 Model Architecture

As shown in Fig. 1, the model architecture consists of different components. We adopt three layers of a multi-layer perceptron network ($\mathbf{N}_{MLP}$) to obtain the embedding of bounding box vectors (v_b) and body pose vector (v_b). We also adopt a 3D convolution network ($\mathbf{N}_{3DC}$) [39] to extract spatio-temporal features of local context region, global semantic, and motion information as follows:

$$
\begin{aligned}
f_b &= \mathbf{N}_{MLP}(v_b \oplus v_p, \mathbf{W}_b), \\
f_{c_l} &= \mathbf{N}_{3DC}(I_l, \mathbf{W}_c), \\
f_{c_g} &= \mathbf{N}_{3DC}(I_g, \mathbf{W}_c), \\
f_{c_o} &= \mathbf{N}_{3DC}(I_o, \mathbf{W}_c),
\end{aligned}
\tag{2}
$$

where $\mathbf{W}_b$ is the weight of the 3-layer MLP, and $\mathbf{W}_c$ is the shared weight of the local, global, and motion features. f_b and $f_{c_i | i \in \{l,g,o\}}$ have the same feature vectors size of 128, where f_b is dedicated to storing the pedestrian location and pose, and f_{cl}, f_{cg}, f_{co} are dedicated to storing the local context information, semantic segmentation, and optical flow (motion features) of the sequential input frames, respectively.

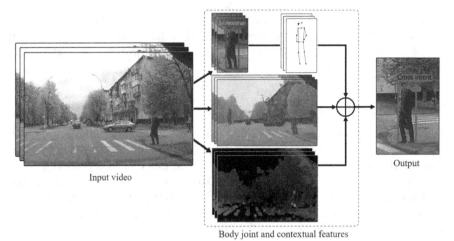

(a) A sample scenario where a pedestrian has an intention to cross the road

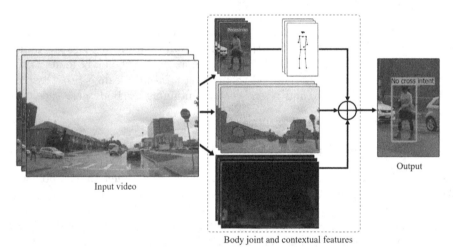

(b) A sample scenario where a pedestrian has no intent to cross the road

Fig. 2. The results of the proposed model in predicting pedestrian intention. Pedestrian local context and body pose, location, the scene motion, and environmental information as global context information are fused to reason about the intention of the pedestrian to cross or not cross the road.

Figure 2 visualises the model flow from input to extracted features followed by crossing intention classification. Our model utilises multiple features; however, the importance factor of each feature might be different for pedestrian crossing classification purposes. Hence, in order to select efficient multi-modal features from a local to a global perspective, the self-attention module [40] (**A**) is applied in each fusion step.

The location and pose feature vector of the pedestrian f_b^t at time t is built based on the extracted temporal attention feature f_A over the past N frames. This helps to understand the importance factor of each feature. Thus, we take the learnable self-attention module, as follows:

$$f_a = \mathbf{A}([f_b^1; ...; f_b^t; ...; f_b^N]), \tag{3}$$

where $[;]$ represents the stack concatenation over temporal dimensions. Afterwards, the fusion of local context features, global context, motion, and the combined location and pose feature vector is specified as:

$$f_H = \mathbf{A}([f_a; \mathbf{A}([f_{c_o}; \mathbf{A}([f_{c_g}; f_{c_l}])])]), \tag{4}$$

Similarly, here $[;]$ denotes the concatenation over the feature modal dimension, and $f_H \in \mathbb{R}^{128 \times K}$, where K denotes the number of feature models and is set as 3 in our fusion strategy. Then, this vector is fed to a global average pooling (GAP) layer to enforce correspondences between feature maps and categories, to be judged by a binary classifier. The classifier consists of two fully-connected layers, and a *Softmax* layer. The *Softmax* layer determines the probability of crossing or not crossing intention within the range of 0 to 1.

4   Experiments

In this section, we discuss the JAAD dataset, the model hyperparameters, and pre-processing settings. We also review the performance of the model compared with similar methods. All experiments are conducted on a PC workstation with an Intel © Xeon W-2225 4.10 GHz processor and dual NVIDIA RTX 5000 GPU with CUDA version 12.0 to benefit the unified shared memory and accelerate the training and inference phases using parallel computing.

4.1   Dataset

The JAAD dataset [11] offers 346 high-resolution videos of autonomous driving in everyday urban areas with explicit crossing behaviour annotation. The length of videos ranges from 1 s to 31 s, and they may contain multiple pedestrians in the scene. The ground truth labels on 2D locations for each frame are provided.

The dataset has two subsets, behavioural data (JAAD$_{beh}$), which contains pedestrians who are crossing (495 samples) or those who are about to cross (191 samples), and all data (JAAD$_{all}$), which has additional pedestrians (2100 samples) with non-crossing actions. To ensure a fair comparison and benchmark, we consider the same training/test splits as per [20].

4.2   Implementation Details

The training data includes 250 videos with a total number of 460 pedestrians on the scenes, and the test data includes 96 videos with a total number of 253 pedestrians in all scenes.

The distribution of labels is roughly balanced throughout the dataset, with 44.30% frames labelled as crossing and 55.70% as non-crossing.

As part of the hyperparameters setting, a learning rate of 5×10^{-7}, 40 epochs, AdamW [41] optimiser, and the batch size $(B) = 2$ were applied. The weight decay of AdamW was deactivated, and a constant decay rate of 10^{-4} in fully-connected layers and a dropout of 0.5 in the attention modules were applied for regularisation and to avoid over-fitting.

As per the previous research and suggestions by [25], a fixed-length sequence of $N = 16$ frames is used for spatio-temporal analysis using the JAAD dataset. Given the pedestrian intention (action) commences at $TTE = 0$ (time to event), the training sequences will be extracted from the past 1–2 s, i.e. $fr_{N-1} \in [TTE - 60, TTE - 30]$. For those videos which contain multiple pedestrians on the scene, the training phase will be repeated multiple times to ensure most of the training samples are utilised.

In terms of data augmentation, random horizontal flip, roll rotation, and colour jittering with a probability of 50% are applied; the same change is performed in all images, bounding boxes, and poses of a sequence.

4.3 Quantitative and Qualitative Evaluation

In our experiments, the proposed model was evaluated against the JAAD ground truth, and compared with other state-of-the-art models summarised in [20].

Table 1 shows the quantitative results of the comparison in terms of area under curve (AUC) and F1 score metrics. AUC represents the ability of our classifier to distinguish between both classes of 'crossing' or 'not-crossing'. For example, a value of 0.5 means that classifier behaviour is equivalent to randomly choosing the class. F1 score is the harmonic mean of precision and recall which are well-known criteria for assessing machine learning-based classifiers.

We examine different combinations of input features as an ablation study in Table 1. These features are depicted in Fig. 2 for two different traffic scenarios. Our baseline method relies on a single feature and uses only bounding box coordinates. It reaches the lowest rate of AUC and F1-score among the other methods. A significant improvement by 6% and 3% in AUC and F1 is achieved by adding the motion features to the baseline method. Adding semantic information to parse the global context of the scene increased the AUC and F1 score by 12.5% with respect to the baseline method. Our initial experiments on extracting global contextual information from the scene indicated that considering more traffic-related objects such as other road users and vehicles results in an increase in intention classification accuracy. For example, using only road segmentation instead of the road and road users segmentation could decrease the AUC by 1.5% and F1 by 1.8%.

The experiments reveal the combination of the pedestrian's bounding boxes coordinates, body joints, spatial features, semantic information, and motion features, leads to the best results on the overall AUC and F1 score combined for both $JAAD_{beh}$ and $JAAD_{all}$. Also, our proposed method could achieve the highest AUC for $JAAD_{beh}$.

Table 1. Performance comparison between the proposed model and other models on the benchmark JAAD dataset. Various variations of our progressive study show the effect of adding local context, global contexts, and pose features. B refers to the bounding box, P is the body pose, L is the local context, G refers to the global contexts, and M refers to motion information. The red and green numbers represent the best and second-best results for each column.

Model	Variant	$JAAD_{beh}$		$JAAD_{all}$	
		AUC	F1	AUC	F1
Static	VGG16	0.52	0.71	0.75	0.55
	ResNet50	0.45	0.54	0.72	0.52
ConvLSTM [42]	VGG16	0.49	0.64	0.57	0.32
	ResNet50	0.55	0.69	0.58	0.33
SingleRNN [11]	GRU	0.54	0.67	0.59	0.34
	LSTM	0.48	0.61	0.75	0.54
I3D [43]	RGB	0.56	0.73	0.74	0.63
	Optical Flow	0.51	0.75	0.80	0.63
C3D [39]	RGB	0.51	0.75	0.81	0.65
ATGC [11]	AlexNet	0.41	0.62	0.62	0.76
MultiRNN [43]	GRU	0.50	0.74	0.79	0.58
StackedRNN [44]	GRU	0.60	0.66	0.79	0.58
HRNN [45]	GRU	0.50	0.63	0.79	0.59
SFRNN [11]	GRU	0.45	0.63	0.84	0.65
TwoStream [46]	VGG16	0.52	0.66	0.69	0.43
PCPA [20]	C3D	0.50	0.71	0.87	0.68
CAPformer [26]	Timesformer	0.55	0.76	0.72	0.55
Spi-Net [30]	Skelton	0.59	0.61	0.71	0.50
TrouSPI-net [47]	Skelton	0.59	0.76	0.56	0.32
Our Method	B	0.50	0.63	0.67	0.51
	B+L	0.51	0.69	0.69	0.57
	B+L+G	0.53	0.75	0.82	0.71
	B+L+G+P	0.52	0.75	0.83	0.72
	B+L+G+P+M	0.60	0.75	0.85	0.73

In addition, the results obtained from our preliminary experiments show that the outcome of the model highly depends on the starting point and moment of the analysis. This is possibly due to the complexity of the environment and also the occasions of partial and fully occluded pedestrians in the scene. This is in line with the claims by [26] and [48]. We also observed that using an accurate pose estimator and scene objects/motion descriptor (e.g., semantic segmentation, optical flow, etc.) leads to outperforming results on the currently available datasets.

Figure 2 illustrates the extracted features among two different video samples of the JAAD dataset and the final prediction of our model in intention classification. The model reports 96% and 87% confidence rates of classification for the samples shown in Fig. 2a and 2b, respectively.

5 Conclusion

We introduced the fusion of various pedestrian and scene features for a better prediction of the pedestrian's crossing intention. The study showed that assessing pedestrian behaviour based on single individual features such as the pedestrian's location, body pose, or global context of the environment will lead to marginally lower performance than the proposed model due to the absence of other complementary influencing factors. It is therefore concluded that relying on multiple features including appearance, pedestrian pose, and vehicle motion, as well as the surrounding information produces better results. We also hypothesised that utilisation of the local and global spatio-temporal features helps to better understand and predict pedestrian crossing intention. Our experiments showed that although the proposed method is not the best from all aspects, it is the superior model overall, achieving the best results in combined AUC and F1-score among 15 other models on JAAD dataset metrics.

The majority of current studies including this research, train their intention prediction model based on the AV's point of view videos. This is while pedestrians may change their decisions also under the influence of AV's speed, distance, lane, and manoeuvre behaviour. In other words, the pedestrians' point of view has been widely neglected. Therefore, the integration of ego-vehicle kinematics into the prediction models can be a sensible approach for future studies.

Acknowledgement. The authors would like to thank all partners within the Hi-Drive project for their cooperation and valuable contribution. This research has received funding from the European Union's Horizon 2020 research and innovation programme, under grant agreement No. 101006664. The article reflects only the authors' view and neither European Commission nor CINEA is responsible for any use that may be made of the information this document contains.

References

1. Ridel, D., Rehder, E., Lauer, M., Stiller, C., Wolf, D.: A literature review on the prediction of pedestrian behavior in urban scenarios. In: 2018 21st International Conference on Intelligent Transportation Systems (ITSC), pp. 3105–3112. IEEE (2018)
2. Tian, K., et al.: Explaining unsafe pedestrian road crossing behaviours using a psychophysics-based gap acceptance model. Saf. Sci. **154**, 105837 (2022)
3. Serpush, F., Rezaei, M.: Complex human action recognition using a hierarchical feature reduction and deep learning-based method. SN Comput. Sci. **2**, 1–15 (2021)
4. Rezaei, M., Azarmi, M., Mir, F.M.P.: 3D-Net: monocular 3D object recognition for traffic monitoring. Expert Syst. Appl. **227**, 120253 (2023)

5. Schneemann, F., Heinemann, P.: Context-based detection of pedestrian crossing intention for autonomous driving in urban environments. In: 2016 IEEE/RSJ International Conference on Intelligent Robots and Systems (IROS), pp. 2243–2248 (2016)
6. Yang, B., et al.: Crossing or not? Context-based recognition of pedestrian crossing intention in the urban environment. IEEE Trans. Intell. Transp. Syst. **23**, 5338–5349 (2021)
7. Sharma, N., Dhiman, C., Indu, S.: Pedestrian intention prediction for autonomous vehicles: a comprehensive survey. Neurocomputing **508**, 120–152 (2022)
8. Wang, J., Huang, H., Li, K., Li, J.: Towards the unified principles for level 5 autonomous vehicles. Engineering **7**(9), 1313–1325 (2021)
9. Kong, Y., Fu, Y.: Human action recognition and prediction: a survey. Int. J. Comput. Vision **130**(5), 1366–1401 (2022)
10. Alahi, A., Goel, K., Ramanathan, V., Robicquet, A., Fei-Fei, L., Savarese, S.: Social LSTM: human trajectory prediction in crowded spaces. In: Proceedings of the IEEE Conference on Computer Vision and Pattern Recognition, pp. 961–971 (2016)
11. Kotseruba, I., Rasouli, A., Tsotsos, J.K.: Do they want to cross? Understanding pedestrian intention for behavior prediction. In: 2020 IEEE Intelligent Vehicles Symposium (IV), pp. 1688–1693. IEEE (2020)
12. Zaidi, S.S.A., Ansari, M.S., Aslam, A., Kanwal, N., Asghar, M., Lee, B.: A survey of modern deep learning based object detection models. Digital Signal Process. **126**, 103514 (2022)
13. Chen, F., Wang, X., Zhao, Y., Lv, S., Niu, X.: Visual object tracking: a survey. Comput. Vis. Image Underst. **222**, 103508 (2022)
14. Cao, J., Pang, Y., Xie, J., Khan, F.S., Shao, L.: From handcrafted to deep features for pedestrian detection: a survey. IEEE Trans. Pattern Anal. Mach. Intell. **44**(9), 4913–4934 (2021)
15. Korbmacher, R., Tordeux, A.: Review of pedestrian trajectory prediction methods: comparing deep learning and knowledge-based approaches. IEEE Trans. Intell. Transp. Syst. (2022)
16. Rasouli, A., Tsotsos, J.K.: Autonomous vehicles that interact with pedestrians: a survey of theory and practice. IEEE Trans. Intell. Transp. Syst. **21**(3), 900–918 (2019)
17. Pop, D.O., Rogozan, A., Chatelain, C., Nashashibi, F., Bensrhair, A.: Multi-task deep learning for pedestrian detection, action recognition and time to cross prediction. IEEE Access **7**, 149318–149327 (2019)
18. Bouhsain, S.A., Saadatnejad, S., Alahi, A.: Pedestrian intention prediction: a multi-task perspective. arXiv preprint arXiv:2010.10270 (2020)
19. Mínguez, R.Q., Alonso, I.P., Fernández-Llorca, D., Sotelo, M.A.: Pedestrian path, pose, and intention prediction through Gaussian process dynamical models and pedestrian activity recognition. IEEE Trans. Intell. Transp. Syst. **20**(5), 1803–1814 (2018)
20. Kotseruba, I., Rasouli, A., Tsotsos, J.K.: Benchmark for evaluating pedestrian action prediction. In: Proceedings of the IEEE/CVF Winter Conference on Applications of Computer Vision, pp. 1258–1268 (2021)
21. Razali, H., Mordan, T., Alahi, A.: Pedestrian intention prediction: a convolutional bottom-up multi-task approach. Transp. Res. Part C Emerg. Technol. **130**, 103259 (2021)

22. Jiang, Y., Han, W., Ye, L., Lu, Y., Liu, B.: Two-stream 3D MobileNetV3 for pedestrians intent prediction based on monocular camera. In: Zhang, H., et al. (eds.) Neural Computing for Advanced Applications: Third International Conference, NCAA 2022, Jinan, China, 8–10 July 2022, Proceedings, Part II, vol. 1638, pp. 247–259. Springer, Cham (2022). https://doi.org/10.1007/978-981-19-6135-9_19

23. Saleh, K., Hossny, M., Nahavandi, S.: Real-time intent prediction of pedestrians for autonomous ground vehicles via spatio-temporal densenet. In: 2019 International Conference on Robotics and Automation (ICRA), pp. 9704–9710. IEEE (2019)

24. Quan, R., Zhu, L., Wu, Y., Yang, Y.: Holistic LSTM for pedestrian trajectory prediction. IEEE Trans. Image Process. **30**, 3229–3239 (2021)

25. Liu, B., et al.: Spatiotemporal relationship reasoning for pedestrian intent prediction. IEEE Robot. Autom. Lett. **5**(2), 3485–3492 (2020)

26. Lorenzo, J., et al.: CAPformer: pedestrian crossing action prediction using transformer. Sensors (Basel, Switzerland) **21**, 5694 (2021)

27. Chen, T., Tian, R., Ding, Z.: Visual reasoning using graph convolutional networks for predicting pedestrian crossing intention. In: Proceedings of the IEEE/CVF International Conference on Computer Vision, pp. 3103–3109 (2021)

28. Zhao, J., Xu, H., Wu, J., Zheng, Y., Liu, H.: Trajectory tracking and prediction of pedestrian's crossing intention using roadside lidar. IET Intel. Transport Syst. **13**(5), 789–795 (2019)

29. Saleh, K., Hossny, M., Nahavandi, S.: Intent prediction of pedestrians via motion trajectories using stacked recurrent neural networks. IEEE Trans. Intell. Veh. **3**(4), 414–424 (2018)

30. Gesnouin, J., Pechberti, S., Bresson, G., Stanciulescu, B., Moutarde, F.: Predicting intentions of pedestrians from 2D skeletal pose sequences with a representation-focused multi-branch deep learning network. Algorithms **13**(12), 331 (2020)

31. Piccoli, F., et al.: FuSSI-Net: fusion of spatio-temporal skeletons for intention prediction network. In: 2020 54th Asilomar Conference on Signals, Systems, and Computers, pp. 68–72. IEEE (2020)

32. Yang, D., Zhang, H., Yurtsever, E., Redmill, K.A., Ozguner, U.: Predicting pedestrian crossing intention with feature fusion and spatio-temporal attention. IEEE Trans. Intell. Veh. **7**, 221–230 (2021)

33. Neogi, S., Hoy, M., Dang, K., Yu, H., Dauwels, J.: Context model for pedestrian intention prediction using factored latent-dynamic conditional random fields. IEEE Trans. Intell. Transp. Syst. **22**(11), 6821–6832 (2020)

34. Neumann, L., Vedaldi, A.: Pedestrian and ego-vehicle trajectory prediction from monocular camera. In: Proceedings of the IEEE/CVF Conference on Computer Vision and Pattern Recognition, pp. 10204–10212 (2021)

35. Singh, A., Suddamalla, U.: Multi-input fusion for practical pedestrian intention prediction. In: Proceedings of the IEEE/CVF International Conference on Computer Vision, pp. 2304–2311 (2021)

36. Wang, W., et al.: InternImage: exploring large-scale vision foundation models with deformable convolutions. arXiv preprint arXiv:2211.05778 (2022)

37. Ilg, E., Mayer, N., Saikia, T., Keuper, M., Dosovitskiy, A., Brox, T.: FlowNet 2.0: evolution of optical flow estimation with deep networks. In: Proceedings of the IEEE Conference on Computer Vision and Pattern Recognition, pp. 2462–2470 (2017)

38. Mordan, T., Cord, M., P'erez, P., Alahi, A.: Detecting 32 pedestrian attributes for autonomous vehicles. IEEE Trans. Intell. Transp. Syst. **23**, 11823–11835 (2020)

39. Tran, D., Bourdev, L., Fergus, R., Torresani, L., Paluri, M.: Learning spatiotemporal features with 3D convolutional networks. In: Proceedings of the IEEE International Conference on Computer Vision, pp. 4489–4497 (2015)

40. Zhao, H., Jia, J., Koltun, V.: Exploring self-attention for image recognition. In: Proceedings of the IEEE/CVF Conference on Computer Vision and Pattern Recognition, pp. 10076–10085 (2020)

41. Loshchilov, I., Hutter, F.: Decoupled weight decay regularization. arXiv preprint arXiv:1711.05101 (2017)

42. Shi, X., Chen, Z., Wang, H., Yeung, D.-Y., Wong, W.-K., Woo, W.-C.: Convolutional LSTM network: a machine learning approach for precipitation nowcasting. In: Advances in Neural Information Processing Systems, vol. 28 (2015)

43. Bhattacharyya, A., Fritz, M., Schiele, B.: Long-term on-board prediction of people in traffic scenes under uncertainty. In: Proceedings of the IEEE Conference on Computer Vision and Pattern Recognition, pp. 4194–4202 (2018)

44. Ng, J.Y.-H., Hausknecht, M., Vijayanarasimhan, S., Vinyals, O., Monga, R., Toderici, G.: Beyond short snippets: deep networks for video classification. In: Proceedings of the IEEE Conference on Computer Vision and Pattern Recognition, pp. 4694–4702 (2015)

45. Du, Y., Wang, W., Wang, L.: Hierarchical recurrent neural network for skeleton based action recognition. In: Proceedings of the IEEE Conference on Computer Vision and Pattern Recognition, pp. 1110–1118 (2015)

46. Simonyan, K., Zisserman, A.: Two-stream convolutional networks for action recognition in videos. In: Advances in Neural Information Processing Systems, vol. 27 (2014)

47. Gesnouin, J., Pechberti, S., Stanciulescu, B., Moutarde, F.: TrouSPI-Net: spatiotemporal attention on parallel atrous convolutions and U-GRUs for skeletal pedestrian crossing prediction. In: 2021 16th IEEE International Conference on Automatic Face and Gesture Recognition (FG 2021), pp. 01–07. IEEE (2021)

48. Gesnouin, J., Pechberti, S., Stanciulescu, B., Moutarde, F.: Assessing cross-dataset generalization of pedestrian crossing predictors. In: 2022 IEEE Intelligent Vehicles Symposium (IV), pp. 419–426. IEEE (2022)

Routes Analysis and Dependency Detection Based on Traffic Volume: A Deep Learning Approach

Maryam Esmaeili$^{(\boxtimes)}$ 🆔 and Ehsan Nazerfard 🆔

Department of Computer Engineering, Amirkabir University of Technology, Tehran, Iran
{Maryamsmaeili,nazerfard}@aut.ac.ir

Abstract. The previous decades have witnessed the remarkable growth of information technology and the emergence of novel algorithms in identifying and predicting future situations. Accordingly, many different methods were proposed in this field. The current paper focuses on two issues. The first detects the movement modes of moving objects in the future based on the current movement route of moving objects. The second calculates the movement dependence degree of moving objects. As a result, the impact of the increase in moving objects is analyzed according to the amount of traffic on the routes. In order to achieve more accurate results, Deep Learning (DL) was used to predict the movement states of moving objects for the future. For this purpose, the raw motion data of moving objects are computed in the Global Positioning System (GPS) format. The extent of route interactions is evaluated by applying new properties to the volume of moving objects and calculating the correlation coefficient and distance criterion, creating a distance matrix for both current and future states. This paper's findings benefit the experts in urban traffic management that can analyze and evaluate the impact of new decisions in advance without spending much time and money.

Keywords: Moving objects · Urban transportation network · Distance criterion · Correlation coefficient · Route prediction · Deep learning

1 Introduction

Traffic volume is regarded as crucial information in many applications like long-range transportation planning and traffic operation analysis. Proposing a practical scheme to capture the traffic volumes on a network scale is useful for Transportation Systems Management & Operations (TSM&O) [1]. Notably, a practical method to prevent freeway incidents is to consider route diversion, which reduces non-recurrent congestion [1, 2]. In recent years, much attention has been devoted to positioning technology and global positioning system (GPS)-enabled devices due to the remarkable development of tracking data [3]. On the other hand, congestion in cities is a severe problem in the modern world, costing billions of hours wasted annually and harming productivity and the global economy [4]. Route prediction can be useful in various situations, including traffic control, expected traffic hazards, and advertising near highways [5].

© The Author(s), under exclusive license to Springer Nature Switzerland AG 2023
M. Ghatee and S. M. Hashemi (Eds.): ICAISV 2023, CCIS 1883, pp. 14–38, 2023.
https://doi.org/10.1007/978-3-031-43763-2_2

The trajectories and movement track are the most popular techniques that characterize a moving object's behavior. The data are represented by three-dimensional (3D) points (x, y, and t), with two spatial dimensions and one temporal dimension [6, 7]. Understanding passengers' route selection behavior is the key to success in predicting future movements in transportation research [8]. Hence, many attempts have been made to forecast travel based on prior data and mobility trends [9]. Endo, Nishida, Toda, and Sawada [10] proposed a method for destination prediction by querying historical data based on RNNs. The authors avoided the data sparsity, and their proposed method could model long-term dependencies. In another study, Long Short-Term Memory (LSTM) recurrent neural networks were proposed by Toqué, Côme, El Mahrsi, and Oukhellou [11] to anticipate dynamic origin-destination (OD) matrices in a subway network. As reported in the related studies, many different data exploration methods have been presented for determining movement patterns so far. Hence, the variation in the distance implies the different characteristics of moving objects which is usually high. Due to these factors, basic requirements of traffic management and related planning can be addressed.

Route prediction can be obtained based on traffic volume for a more accurate and complete prediction. Massive moving objects with continuous changes in location create a large volume of data. These enormous spatiotemporal data of moving objects must be collected and stored [12]. Due to the influx of data, acquiring moving object databases (MODBs) and preprocessing them is crucial. In this step, the data are registered and stored by GPS recorders to represent the current state of the urban network. Then, the dataset is converted into an accurate dataset based on the primary information and standard form of geographical coordinates. For this purpose, the maps' matching method is commonly adopted, which can overlap the points outside the main route based on the noise of GPS devices and turn them into the main points of the desired route. This method creates a new data source according to the standard roadmap. Once the data are gathered, a new dataset must be developed to indicate the distance criterion for the whole route according to the research objectives. The distance criterion demonstrates the extent to which routes influence one another. The distance matrix can effectively specify the distance between the routes that can be connected to others. The distance matrices were extensively examined in biology and anthropology, as mentioned in [13–15]. Also, there are many additional and more recent references regarding distance matrices in public transport [16, 17]. In the current study, a distance matrix calculates the distance criterion for the various routes. Accordingly, the distance matrix is used for the current state of the roadmap based on the volume of route traffic.

Besides, a new dataset is obtained for the next or future state. The selected DL algorithms that solve the considered problem are LSTM and Convolutional Neural Networks (CNN), based on which the route requires a series of the previous ones. The routes considered here are related to the public transport vehicles in a traffic system. As an innovation, a novel distance matrix is calculated to obtain current and future datasets. Hence, the relationship between the components of both matrices is determined. The correlation coefficient is also considered for calculating the relationship between the distance criteria of the current and future states. As a result, any change or limitation in traffic volume at any point on the roadmap is evaluated before being imposed. The major aim is to make the best decision without wasting time and money. Urban network analysis can help drivers improve roads by introducing efficient resources and facilitating traffic congestion in urban areas [18].

Despite tremendous interest in this regard, many gaps and shortcomings need to be tackled. The previous studies succeeded in offering different methods for route prediction and still compete with each other regarding accuracy and authenticity. A few studies considered the whole factors, including traffic volume, weather conditions, and so on, for a more accurate prediction. Computer-aided techniques like Machine Learning (ML) and DL models contribute to predicting the route. This research employs DL algorithms based on traffic volume for route prediction. As a novelty, the traffic volume information is determined at a specific time for obtaining the two distance matrices in the present and the future. In order to compare the current and future trends, the correlation coefficient is considered. The proposed method can be on par with many state-of-art prediction models and even outperforms them in accuracy and authenticity.

The rest of this paper is organized as follows: Sect. 2 reviews the related studies to highlight the main gaps. In Sect. 3, we present the methods that assist us in reaching our goal. Section 4 illustrates the proposed method in detail based on the research objectives. The experiments and results are given in Sect. 5. The conclusions are drawn, and future studies are suggested in Sect. 6.

2   Literature Review

Ashbrook and Starner [19] offered historical GPS data for user destination prediction. This method is split into two parts: (1) clustering the raw GPS coordinates into candidate destinations and (2) predicting the next destination using a Markov model via the candidate locations as states. Several researchers have employed this approach Alvarez-Garcia, Ortega, Gonzalez-Abril, and Velasco [20], Panahandeh [21], Simmons, Browning, Zhang, and Sadekar [22], Zong, Tian, He, Tang and Lv [23]. Kamble and Kounte [24] presented a method for route prediction using a clustering algorithm and their GPS sensor data. Marmasse and Schmandt [25] experimented with a Bayes classifier, histogram matching, and a hidden Markov model to match a partial route with stored routes. Krumm [26] utilized a Markov model to quickly anticipate a driver's direction. A hidden Markov model was employed by Simmons, Browning, Zhang, and Sadekar [22] to forecast destinations and itineraries. By examining vehicle routes compiled by 250 drivers, Froehlich and Krumm [27] determined route regularity, and the closest match algorithm subsequently yielded an ordered list of route candidates based on route regularity.

Furthermore, Liu, Wu, Wang, and Tan [28] utilized an RNN-based model based on check-in history to forecast the future location. However, based on previous data, their model could be more appropriate for multi-step prediction. In other words, whereas check-in data may be easily computed by considering a single transition, trajectory data must evaluate transitions between many location points on the pathways from the current position to a destination. Laasonen [29] proposed route prediction using string-matching algorithms to a stored route database. Epperlein, Monteil, Liu, Gu, Zhuk, and Shorten [30] introduced a model based on Markov chains to probabilistically anticipate the route of the current journey and form a Bayesian framework to model route patterns. Vahedian, Zhou, Tong, Li, and Luo [31] also predicted event gathering using trip destination prediction from taxi data. De Sousa, Boukerche, and Loureiro [32] suggested a cluster-based method for the long-term prediction of road-network restricted trajectories. De Brébisson, Simon, Auvolat, Vincent, and Bengio [33] also used a multilayer perceptron (MLP) to tackle the destination prediction problem as a regression problem. Their method is based on the idea that the location of a destination can be represented as a linearly weighted combination of famous destination clusters, with the weights determined using the MLP. Ke, Zheng, Yang, and Chen [34] noted that the DL strategy might better capture the spatiotemporal properties and correlations of explanatory variables due to merging convolutional techniques with the LSTM network. Zhao, Zhao, and Cui [35] proposed a new framework for measuring network centrality that considers a road network's topological and geometric aspects. They acquired the correlation coefficient of Wuhan, China, GPS data to test their approach. Choi, Yeo, and Kim [36] suggested a deep learning approach for learning and predicting network-wide vehicle movement patterns in metropolitan networks that can replicate real-world traffic patterns. Dai, Ma, and Xu [37] used a deep learning framework, performed spatiotemporal correlation analysis of traffic flow data, and obtained a space-time correlation matrix containing traffic flow information for short-term traffic flow prediction. Gao, Wang, Gao, and Liu [38] estimated urban traffic flow in Qingdao, China, using GPS-enabled taxi trajectory data and a weighted correlation coefficient. In the study by Huang, Deng, Wan, Mi, and Liu [39], the traffic states in the Chongqing Nan'an district were evaluated by Pearson correlation analysis. Finally, Yang, Jia, Qin, Han, and Dong [40] concentrated on the spatiotemporal correlations between different pairs of traffic series.

Moreover, more advances have been observed in recent years regarding the DL and ML models proposed for predicting traffic volume [17]. To mention a few, a novel hybrid boosted long short-term memory ensemble (BLSTME), and CNN was introduced for solving the congestion problem in the vehicular ad-hoc network (VANET) Kothai, Poovammal, Dhiman, Ramana, Sharma, AlZain, Gaba and Masud [41]. Lohrasbinasab, Shahraki, Taherkordi, and Delia Jurcut [42] proposed a novel model, Network Traffic Monitoring and Analysis (NTMA), for solving the problem of the volume of transferred data in large-scale, heterogeneous, and complicated networks. The proposed model was also useful for Network Traffic Prediction (NTP), and its operation was based on the recent relevant works. In 2020, Ke et al. [43] presented a two-stream multi-channel convolutional neural network (TM-CNN) model for predicting the multi-lane traffic speed based on the traffic volume impact. In a review of 2022, many DL algorithms in public transportation systems were examined comprehensively [44]. In another research, Tsai et al. attempted to solve the prediction problem of the number of passengers on a bus [45]. Accordingly, the authors used simulated annealing (SA) to specify an appropriate number of neurons for each layer of a fully connected DNN to improve the accuracy rate in solving the specific optimization problem.

The process of calculating the amount of congestion and the number of cars using a specific piece of road at a specific moment is known as traffic volume. This time frame may be expressed in "minutes," "hours," "days," etc. The previous research mainly needed an accurate distance matrix to obtain current and future datasets. Many service systems, including delivery, customer pick-up, repair, and maintenance services, require a fleet of vehicles initially stationed at a set of depots to satisfy the client's needs. The goal is to identify an inventory of routes that satisfy various requirements and have a small overall length. The issue, sometimes the VRP, was thoroughly examined in recent journals [46]. It is unlikely that a polynomial time algorithm will be created for the problem's ideal solution because it is NP-hard. As a result, heuristic algorithm development has received much attention. The goal and the issue with vehicle routing are to choose the most suitable paths for a fleet of vehicles within operational constraints, such as time window and route length. It aids fleet managers in creating routes that would increase fleet productivity and cut down on last-mile delivery expenses. Notably, the competition among the studies regarding route prediction based on traffic has yet to be over. In the current research, the traffic volume information is calculated at a characteristic time for obtaining the two distance matrices in the present and the future. The proposed model can be on par with the other state-of-art ones and even outperforms them in terms of accuracy and authenticity.

3   Methods

Based on the proposed method, the data that indicate the condition of the road network in the future is required. Moreover, the history of at least four previous routes should be considered, i.e., the previous state versus the next state or the current state versus the future state. The best choice is deep learning algorithms which are CNN and LSTM, in this research. ML is the knowledge that has changed life and the business world to a great extent in our time. Among these transformations, DL is considered a revolutionary method. If ML can make machines work as well as humans, DL is a tool in the hands of humans that can do things better than them. The routes selected in the current research are the public transport vehicles in a traffic system. Notably, a distance matrix is computed to specify the current and future datasets. Then, the relationship between the components of both matrices is calculated. Besides, the correlation coefficient specifies the relationship between the distance criteria of the current and future states. The present section outlines the use of DL techniques considered here.

3.1   CNN

Similar to traditional ANNs, convolutional neural networks (CNNs) are composed of neurons that can adapt to their environment. The next step for each neuron is to receive input and carry out an action (like a scalar). Convolutional layers, layers for pooling, & fully-connected layers are the three types of layers that make up CNNs. Comparatively speaking, CNNs require less preprocessing than other classification algorithms. The filters used in the original approach, however, are manually created. With sufficient practice, CNN might acquire the filters or functions. CNN's architecture was influenced by how the brain's "Visual Cortex" was organized and remained the same as how neurons link in the human brain. Only a small portion of the visual field, referred to as the "receptive field," is used by each neuron to react to stimuli [47].

3.2   LSTM

LSTM (long short-term memory network) is a recursive neural network employed in deep learning because great architecture can be taught successfully. LSTM networks have become an essential tool in deep learning. Many people consider them a suitable alternative to recurrent neural networks. The rapid growth of machine learning research has led to new methods coming out quickly, but long-short-term memory networks have declined. This model is also reliable for long-term goals. It has been used to process and predict time series data and important events with relatively long delays in many fields [48].

3.3   Proposed Model

The method for analyzing the routes and detecting dependency based on traffic volume is outlined here. Figure 1 highlights the proposed framework in which GPS data is collected from the beginning and then mapped. The DL techniques, namely CNN and LSTM, are considered for predicting routes based on traffic volume. The proposed model evaluates the degree of dependence of the routes on each other from the traffic perspective, which can obtain vital information for urban traffic management in the future. Notably, the correlation coefficient is a good measure to indicate the relationship's intensity and the type of relationship (inverse and direct).

Moreover, a typical example can be explained to clarify the proposed model expensively. Accordingly, a bridge or a shopping center is imagined on the street. Now, the impact of this bridge on the volume of traffic and other routes will be examined. The proposed framework obtains the traffic information on the future routes in this case. The X route currently reduces the volume of traffic on the Y route, or vice versa, or it has no effect. The collected assessment supplies the people with vital information that has numerous uses in urban traffic management to prevent traffic congestion, predict trip times, design energy management systems such as hybrid automobiles, etc. Regarding Fig. 1, the increasing number of moving objects and urban constructions has caused a high traffic volume, which will affect the route in other directions. More information regarding the major components of this model is given in the following. The primary aim here is to benefit from computer-aided techniques for enhancing the network and route planning and vehicle and crew scheduling. The aim is achieved when the proposed method's outcomes align with the route analysis and dependency detection based on traffic volume. The findings relate to public transportation and can be generalized to other transportation systems, monitoring, and management.

3.4   Describing the Route of the Road Network Using Graph Theory

It should be noted that the graph theory was adopted to describe the routes network. Hence, there are points on the map in the sense of a node representing the intersection between streets and the endpoint of a street. These routes connect the points, considered edges representing the connections between these nodes. For the sake of simplicity, a directionally connected graph was also considered. Nevertheless, a route type attribute that indicates a one-way and two-way route was utilized in the edge properties. A set

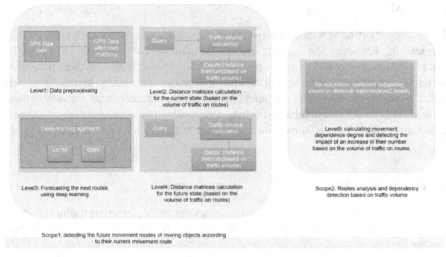

Fig. 1. The structure of the proposed method

of GPS-related data collected by moving objects that can contain information such as geographic coordinates, weather, time, and date. These data are provided in a standard format.

When a navigation system locates the car on a map, map matching is used. In many applications, data like speed restrictions are given to the representations of roads in a digital map; mapping such data to moving objects is known as map matching. Much attention must also be devoted to noisy data recorded by GPS devices. Before processing the data, a preprocessing step was carried out using the map-matching method to indicate points on the main and standard routes. It causes the route data from GPS devices to record data outside the main route, which can be used as the correct input data for the analysis section. Using the map matching method, route data, which consists of the data of GPS devices and has nodes in some routes, can record data outside the main route. Accordingly, the data outside the standard route enter the main route, and subsequently, the correct data are utilized for the input of the analysis section. A considerable body of literature exists on the map-matching method [49, 50]. For instance, a DNN was presented for extracting the spatial features of traffic considering graph convolution and its temporal features based on the LSTM for making short-term and long-term predictions [51]. The authors used the map matching method and proved the capabilities of the proposed method in dealing with the time horizons from 5 min to up to 4 h with multi-step predictions. Zhao et al. employed a GPS map-matching algorithm to present the accuracy and efficiency requirements of matching. The purpose was to predict the traffic speed under non-recurrent congestion [52].

Regarding the recorded coordinate point, the moving objects can record many of the points according to their trajectory. Analyzing these points results in routes taken by the desired moving object. It is even possible to obtain computable data on time, speed, and distance to the origin, which can be important in the discussion of trajectory analysis. Furthermore, the offline method is used to analyze GPS data.

In the proposed method, several data structures are considered for processing and analyzing trajectories and points, which are defined as follows:

Route data structure: To define the routes in the proposed method, we have considered the characteristics of the route ID, longitude and latitude coordinates at the beginning of the route, the source node ID, the destination node ID, and the traffic volume on the route.

Node structure: To define the node, the node ID properties, as well as node longitude and latitude coordinates, are considered in the proposed method.

Moving object structure: To define a moving object in the proposed method, the properties of the registration ID, the moving object ID, the registration period ID, the coordinates of the length and width of the registration point, the registration date, the registration time, and the distance of the route to the destination are examined.

Moving object's trajectory structure: A list of moving object structures is used in the proposed method to define a moving object's trajectory. Each current position record adds a moving object data type to the trajectory list to assess moving objects' motion and positions.

3.5 Definition of the Database Related to the Amount of Traffic on the Routes

Each point recorded by the GPS device has a time dimension as well. In addition to the geographical coordination, the recording time is considered. Accordingly, the time and place dimensions are calculated using the recorded data in the database to obtain the amount of traffic volume for a specific time in each route. Also, the queries are employed to attain the current traffic of the whole route, the traffic of a particular route, the trajectory, speed, the distance of the moving object, and the closest node to the current point of the moving object. Notably, there is significant traffic data considered as the traffic volume.

The current GPS points correspond to the nearest point, indicating the main route based on which the moving object is identified. The query is based on the set time, which identifies the moving objects on the present route X. The query's result indicates the same number of moving objects on the route, which is the so-called traffic volume, depending on the characteristics. The route, the space a moving object can hold, is determined. The moving object's capacity contributes to designing and constructing map routes based on an urban plan for each target volume. In other words, the target volume of moving objects plays a prominent role in design and construction. Also, each route has been specified at a certain length and width by considering an average value of the size of moving objects. The approximate capacity is achieved for obtaining the traffic volume as a percentage between the routes (See Formula 1):

$$D_{AB} = the\ distance\ between\ points\ A\ and\ B$$

$$C_{AB}^{T} = the\ number\ of\ vehicles\ between\ points\ A\ and\ B\ at\ time\ T$$

$$L = the\ length\ of\ vehicles\ and\ the\ safe\ distance\ between\ them$$

$$P = the\ rate\ of\ the\ traffic\ volume\ in\ percentage$$

$$P = \frac{C_{AT}^T \times L}{D_{AB}} \times 100 \tag{1}$$

The distance between the traffic volumes is specified on the various routes according to the Euclidean, Manhattan, and Minkowski distances.

The rows and columns considered in the three types of distance matrix are nodes. As the traffic volume characteristics are considered, the matrix entries represent the traffic volume as a percentage between the routes. Different methods for distance measurement are determined according to the type and number of features. For instance, to determine the distance of a coordinate point, which has two dimensions, both the x and y axes must be considered in calculating the distance between two points. The Manhattan matrix is provided as an example. In this method, only the traffic volume is considered, and as there is only one feature, it is one-dimensional. Based on this, the distance is specified using only one axis. For this purpose, the Manhattan matrix obtains the matrix roots.

$P_1 = x_1$, $P_2 = x_2 \rightarrow$ Manhattan Distance $= |x_1 - x_2|$

The two points, P_1 and P_2, have the characteristics of x_1 and x_2, and the Manhattan distance is obtained from the difference between the two points along the X-axis. Notably, the amount of traffic volume is calculated for different routes according to a specific time. For instance, X = [[2], [5], [6], [7], [9]], which means route 0 has two traffic volumes, route 1 has five traffic volumes, route 2 has six traffic volumes, and so on.

3.6 Forecasting the Next Routes Using Deep Learning

In the proposed method, the available data is preprocessed and analyzed. Then, a distance matrix is obtained from the traffic data of the route network. Since the aim here is to analyze the effects of increasing the volume of moving objects on one route to other routes, a new data source is defined for new or future conditions. The DL algorithms assist in calculating the effect of moving objects and the traffic volume with the highest accuracy. In this step, the moving objects at a point are substituted on the route and considered at least four nodes identical to the current route. A time series of route sequences is defined as deep learning input. DL algorithm determines the next node or route to use the following routes or nodes to complete future ones. Selecting any new route requires a four-member sequence of the last nodes traversed by the moving object. As the time series of the route moves forward, another feature, namely traffic volume, is added to the selected routes. In the following step, the whole routes obtained from the last one in the selected route sequence are selected as a candidate route which can be any number. There are two methods for choosing the next route (the fifth route). If the number of available routes is only one, then the same route will be the final selected route; otherwise, the test result of deep learning will be the ultimate route. An example of a time series of routes considered as an input to deep learning is:

$$[[1\ 2][2\ 3][3\ 3][4\ 2][5\ 5]]$$

In this input, the first four routes are considered as the current model, and the fifth route is the candidate. In other words, the moving object was initially at point 1, while the traffic volume was at 2. Then, it went to point 2 with traffic volume 3; on the next route, it went to point 3 with traffic characteristic 3; finally, it went to point 4 with traffic volume 2. After navigating these routes, the moving object selects candidate route 5 with the traffic volume 5. In this case, a result or input category is binary for the above input. If the value is 1, the above route is selected; otherwise, it is not selected, and value 0 is considered. Collecting these data provides a data source for implementing the deep learning algorithm. By creating this data source, it is now possible to dynamically select the next route from the candidate routes for each moving object route sequence. For example, a moving object traverses routes 1, 2, 4, and 5. To determine the following route, it is assumed that there are five candidate routes. Accordingly, deep learning is applied to the sequence of four routes and the new candidate routes (5). The result of the algorithm is zero or one (0 or 1). Therefore, each of the input sequences that receives the value of one from deep learning can be selected. Our method selects the route with the least traffic volume from the results.

The sample test and results are presented as follows:

$$X = [[1\,2][2\,3][3\,3][4\,2][5\,5]] \rightarrow Deep\,Learning \rightarrow Y = 1$$

Since the time difference between the present and future is constant, the first four routes of the present state must be selected for identifying the routes of the future state. The times considered for these four initial routes in the future state have a constant difference with the passage time of the current state routes. However, for new routes, it can be a variable according to the speed. A specific time is considered for the current and the future states, and the volume of moving objects is dynamically calculated simultaneously for different routes. The volume of moving objects means how many moving objects exist on the route, e.g., 14324232, which is obtained considering the coordinates of the points along with the time dimension. A model can be formed on CNN and LSTM using the processor in the following.

3.7 Analyzing Distance Matrices and Determining the Impact Rate

In order to evaluate, compare, and analyze the impacts of future routes on the current ones, the two current and future states must be equated in distance matrices. Thus, the common routes in the future and present are analyzed. Besides, two distance matrices are considered with the exact dimensions, one for the current and the other for the future. The degree of correlation between the two matrices is calculated according to the Pearson correlation coefficient for obtaining the degree of effectiveness in the analyzed routes. The Formula of the Pearson correlation coefficient is given in Formula 2 in which the values of the x-variable in a sample, the meaning of the values of the x-variable, the values of the y-variable in a sample, the mean of the values of the y-variable, and correlation coefficient are captured by x_i, $\bar{x}$, y_i, $\bar{y}$, and r respectively.

$$r = \frac{\sum(x_i - \bar{x})(y_i - \bar{y})}{\sqrt{\sum(x_i - \bar{x})^2(y_i - \bar{y})^2}} \tag{2}$$

According to the proposed model, the correlation coefficient has the following features:

If the correlation coefficient of the two parameters is zero, it means they are independent of each other. With the available information, it is impossible to comment on the increase or decrease of one parameter compared to the other.

If the correlation coefficient of the two parameters is positive, increasing one parameter increases the other parameter, or decreasing one reduces the other one. In our proposed method, if the correlation coefficient of two routes is positive, it means that the increase of traffic in one route has a positive impact on the other route and can increase the traffic on that route, too.

If the correlation coefficient of two parameters is negative, increasing one parameter decreases the other and vice versa. In our proposed method, if the correlation coefficient of the two routes is negative, increasing the traffic in one route reduces the traffic in the other route and vice versa.

4 Experiments and Results

4.1 Analysis and Investigation

Figure 2 shows the region considered here, located in Los Angeles, California. The behavior of 1000 moving objects, processed based on the spatiotemporal data of moving objects recorded by GPS tools in Extensible Markup Language (XML) format, is analyzed here. A moving object in each journey moves from the origin and stops at the destination. It passes through different nodes and routes to reach the destination. During this time, the GPS records the coordinates at a specific time.

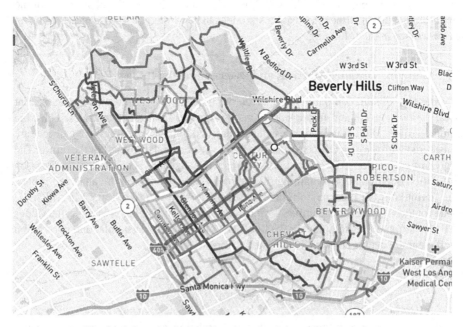

Fig. 2. Selected area for the case study of the proposed method

Considering the proposed method, the data obtained from GPS devices included spatial noise. Standard maps, such as Google Maps, define the routes accurately. For instance, route x contains points that indicate the route through which map-matching operations can be performed. Thus, it is straightforward to determine which route this point belongs to. Accordingly, a sample of the data is given in the map-matching results. Applying the map matching method yields the graph characteristics describing the roadmap and the data source describing the moving objects and their trajectory. Unnecessary nodes and edges are also removed by applying map matching, as depicted in Fig. 3. Due to the change in GPS data points created by the map matching method, a new route data source is formed. The data conform to the standard identifier of the global systems associated with online maps, an example of this data source.

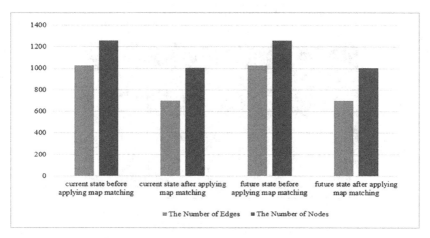

Fig. 3. Graph of road network volume change rate

The traffic on the routes is examined by analyzing the obtained graph. Consequently, the distance matrix from the current state is obtained. The results obtained from the selected methods for calculating the distance matrix, including Euclidean, Manhattan, and Minkowski distance, are presented in Fig. 4. As seen from Fig. 4, the amount of traffic is normalized as distance matrices and has a value between zero and one, which is the criterion of a difference between the percentages of traffic routes. Rows and columns of the matrices indicate the nodes and entries of the matrices indicate the traffic volume on the route displayed as a percentage.

According to Figures 4 and 5, the traffic volume in the paths indicates that the traffic volume in the x path is 20%, whose 20% is occupied. We need future data for making predictions and obtaining the traffic volume in the future paths in each array. For example, the y path in the future traffic volume is 20, meaning it occupies 20% of the path. LSTM and CNN were used to prepare future data. The following route was determined by identifying the first four routes in the proposed deep learning method. The new routes were identified initially, and the traffic volume was computed. The distance matrices were calculated for the future state, as shown in Fig. 5.

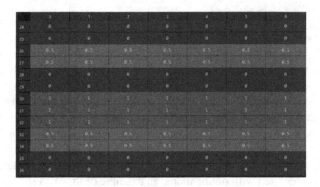

a) Minkowski distance

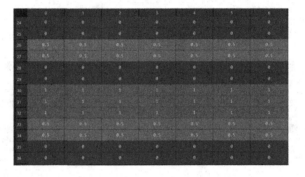

b) Manhattan distance

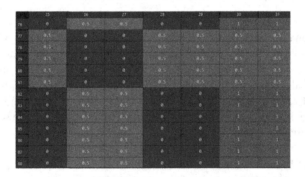

c) Euclidean distance

Fig. 4. Distance matrices calculated for the current state of the roadmap

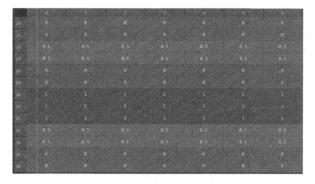

a)Minkowski distance

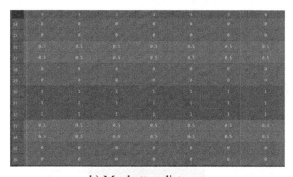

b) Manhattan distance

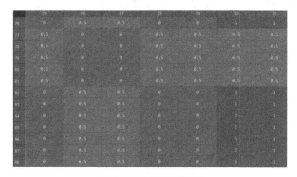

c) Euclidean distance

Fig. 5. Distance matrices calculated for the future state of the roadmap

A schematic view of the CNN and LSTM layers is shown in Fig. 6 and 7, respectively. The input and output for CNN and LSTM are the same. In the previous steps, the data (a GPS file) were preprocessed, and after completing these steps, a time series of our data, route ID, and traffic volume, which were given as the input, was obtained. None means that if 20 samples of the object's movement are given, none will be 20. The output was zero and one. Figure 6 shows a one-dimensional convolution layer, a max pooling layer, and a flattened layer. The number of inputs is equal to the flattened columns. Each of these columns, along with the bias, goes to the output of the dense layer network. In Fig. 7, in the LSTM layer, there exist five cells, and the output is none in one hundred. It goes to the dense layer, to which the Sigmoid function is dedicated.

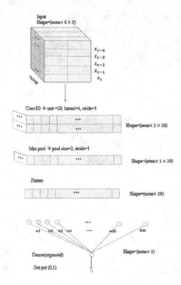

Fig. 6. Schematic view of CNN layers for route prediction based on traffic volume in urban transportation networks

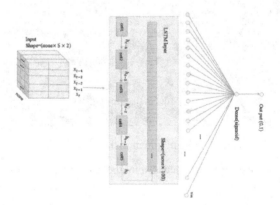

Fig. 7. Schematic view of LSTM layers for route prediction based on traffic volume in urban transportation networks

Notably, two distance matrices related to the current and future states can evaluate the effect of changes in the number of moving objects at any point or route on another route. The correlation coefficient test result related to the two mentioned matrices is displayed in Fig. 8. Also, the effect of changing each route on other routes is examined. Accordingly, it was revealed how the correlation coefficient or the changes of each route affect others based on the traffic volume. This method can save time and money in many urban network decisions or operations. The impact of an increase in volume is significant by applying different loads on different routes. Hence, using the proposed method, money and time are effectively saved when making urban network decisions.

In addition, Fig. 9 shows the 3D diagram of the routes and the correlation coefficient. Accordingly, a rise in traffic volume is observed in yellow and gray, while the decrease is displayed in orange and blue. Due to the obtained results, the benefits of the correlation coefficient for the two distance matrices are obvious. Hence, the type and value of the correlation between the main parameters are adequately characterized. The correlation coefficient contributes to analyzing distance matrices and determining the impact of future routes on the current ones.

4.2 Comparing CNN and LSTM Models

Loss and accuracy functions of LSTM are depicted respectively in Fig. 10(a) and 10(b). Similarly, the loss and accuracy functions of CNN are shown respectively in Fig. 10(c) and 10(d) to compare CNN and LSTM based on our objective. As expected, the loss function gradually decreased with more epochs, which means our model was trained. In both CNN and LSTM methods, the loss function was somewhat similar. It can be claimed to perform better in the CNN method because it is more stable. LSTM achieved an accuracy of 0.9300860583782196, while CNN achieved an accuracy of 0.911368329524993.

4.3 Comparing with the Related Works

As seen from the literature, more studies have been conducted regarding this research topic. In 2022, Ganapathy et al. predicted highway traffic flow according to spatial-temporal traffic sequences for various origin-destination (OD) pairs [53]. Through extensive experimentation on the actual traffic network of Chennai Metropolitan City, the performance of Spatial-Temporal Reconnec (STAR) was examined. An empirical analysis was conducted considering the algorithm's computational complexity. Compared to previous baseline approaches in short-term traffic flow predictions like LSTM, ConvLSTM, and GRNN, the suggested STAR algorithm predicted traffic flow during peak hour traffic with lower computational cost. In another work, a hybrid model was proposed for air traffic dependence which incorporates fuzzy c-means (FCM) and graph convolution network (GCN) [54]. Through clustering trajectory data using FCM, practice demonstrated that the ATFN graph structure might accurately capture crossing sites and primary pathways. The GCN was used to visualize the output characteristic layer, which can efficiently identify the temporal and spatial correlation of the flow state. In order to detect the trajectory deviation, time delay, and tolerance of the ATFN at different times or paths, the proposed method was considered. According to experimental findings using actual data, FCM-GCN performed much better in mid- and long-term traffic

	27	28	29	30	31	32	33
56	0.67452	-1	-1	1	1	1	0.67452
57	0.67452	-1	-1	1	1	1	0.67452
58	1	-0.67452	0.67452	0.67452	0.67452	0.67452	1
59	-0.67452	1	1	-1	-1	-1	-0.67452
60	-0.67452	1	1	-1	-1	-1	-0.67452
61	1	0.67452	0.67452	0.67452	0.67452	0.67452	1
62	1	-0.67452	-0.67452	0.67452	0.67452	0.67452	1
63	-0.67452	1	1	-1	-1	-1	-0.67452
64	1	-0.67452	-0.67452	0.67452	0.67452	0.67452	1
65	-0.67452	1	1	1	-1	-1	0.67452
66	-0.67452	1	1	-1	-1	-1	-0.67452
67	-0.67452	1	1	-1	-1	-1	-0.67452
68	-0.67452	1	1	-1	-1	-1	-0.67452
69	-0.67452	1	1	-1	-1	-1	-0.67452

a) Euclidean distance

	1365	1366	1367	1368	1369	1370	1371
51	0.85831	0.85831	0.85831	0.85831	0.85831	0.85831	0.85831
52	0.85831	0.85831	0.85831	0.85831	0.85831	0.85831	0.85831
53	0.576464	-0.576464	0.576464	0.576464	-0.576464	-0.576464	0.576464
54	-0.85831	-0.85831	0.85831	-0.85831	-0.85831	0.85831	-0.85831
55	-0.85831	-0.85831	-0.85831	-0.85831	-0.85831	-0.85831	-0.85831
56	-0.85831	-0.85831	-0.85831	-0.85831	-0.85831	-0.85831	-0.85831
57	0.85831	0.85831	-0.85831	-0.85831	0.85831	-0.85831	-0.85831
58	-0.576464	-0.576464	-0.576464	-0.576464	-0.576464	-0.576464	-0.576464
59	0.85831	0.85831	0.85831	0.85831	0.85831	0.85831	0.85831
60	0.85831	0.85831	0.85831	0.85831	0.85831	0.85831	0.85831
61	-0.576464	-0.576464	-0.576464	-0.576464	-0.576464	-0.576464	-0.576464
62	-0.576464	-0.576464	-0.576464	-0.576464	-0.576464	-0.576464	-0.576464
63	0.85831	0.85831	0.85831	0.85831	0.85831	0.85831	0.85831
64	-0.576464	-0.576464	-0.576464	-0.576464	-0.576464	-0.576464	-0.576464

b) Minkowski distance

	27	28	29	30	31	32	33
56	0.67452	-1	-1	1	1	1	0.67452
57	0.67452	-1	-1	1	1	1	0.67452
58	1	-0.67452	-0.67452	0.67452	0.67452	0.67452	1
59	-0.67452	1	1	-1	-1	-1	-0.67452
60	-0.67452	1	1	-1	-1	-1	-0.67452
61	1	-0.67452	-0.67452	0.67452	0.67452	0.67452	1
62	1	-0.67452	0.67452	0.67452	0.67452	0.67452	1
63	-0.67452	1	1	-1	-1	-1	-0.67452
64	1	-0.67452	-0.67452	0.67452	0.67452	0.67452	1
65	-0.67452	1	1	-1	-1	1	-0.67452
66	-0.67452	1	1	-1	1	1	0.67452
67	-0.67452	1	1	-1	-1	1	-0.67452
68	-0.67452	1	1	-1	-1	-1	-0.67452
69	-0.67452	1	1	-1	-1	-1	-0.67452

c) Manhattan distance

Fig. 8. Result of calculating the correlation coefficient based on distance matrices

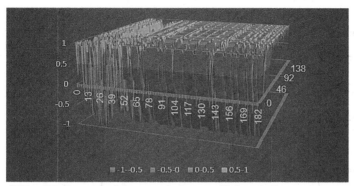

a) The 3D diagram of the correlation coefficient in the current state

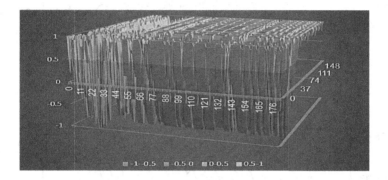

b) Diagram of the correlation coefficient rate in the future

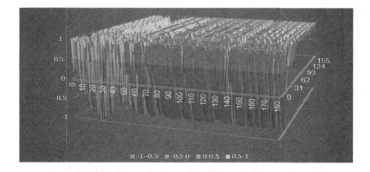

c) 3D diagram of the correlation coefficient in the current and future states

Fig. 9. The correlation coefficient of distance matrices in the current and future states (Two dimensions are the route, and the other is the correlation coefficient.)

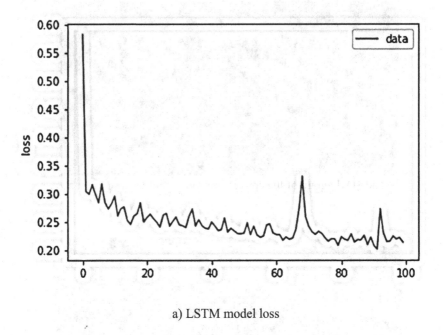

a) LSTM model loss

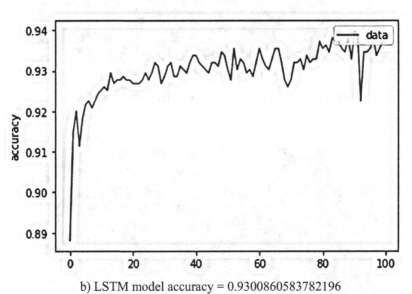

b) LSTM model accuracy = 0.9300860583782196

Fig. 10. Loss function and accuracy in CNN and LSTM

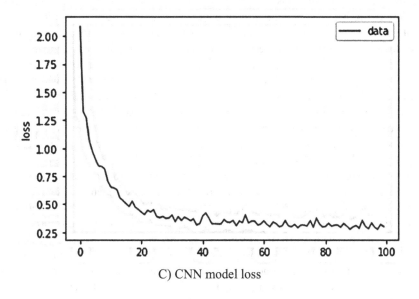

C) CNN model loss

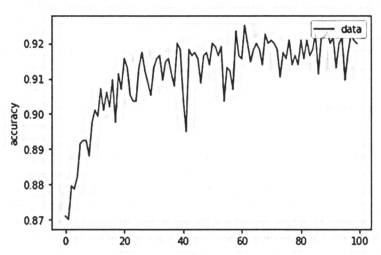

d) CNN model accuracy = 0.9113683295249939

Fig. 10. (*continued*)

forecasting tasks than other baseline models. The traffic order was analyzed in another
work employing ML and a multi-source data set, including traffic conditions, traffic
control devices, road conditions, and external factors. Data were gathered from certain
Beijing highway exits using field research and navigation. As a new surrogate index, the
traffic order index (TOI), which depended on aggregate driving behavior data, was used
to assess the safety risk. Also, eXtreme Gradient Boosting (XGBoost) ML was modified
to create a traffic order prediction model. In order to forecast traffic flows using data
from radar and weather sensors, Hang et al. compared various ML techniques, including
long short-term memory, autoregressive LSTM, a convolutional neural network, and
data properties [55]. Loss and accuracy functions of LSTM are depicted respectively
in Fig. 10(a) and 10(b). Similarly, the loss and accuracy functions of CNN are shown
respectively in Fig. 10(c) and 10(d) to compare CNN and LSTM based on our objective.
As expected, the loss function gradually decreased with more epochs, which means our
model was trained. In both CNN and LSTM methods, the loss function was somewhat
similar. It can be claimed to perform better in the CNN method because it is more stable.
LSTM achieved an accuracy of 0.9300860583782196, while CNN achieved an accuracy
of 0.911368329524993. The superiority of the proposed model over the previous ones
lies in the heart of the accuracy and authenticity of the outcomes. DL algorithms determined by traffic volume were used for route prediction in this study. The two distance
matrices for the present and the future were obtained, for the first time, using information
on traffic volume determined at a certain time. The correlation coefficient was considered when comparing the present and upcoming developments. The suggested approach
could compete with numerous cutting-edge prediction models and even beat them in
precision and veracity.

5   Conclusion

In summary, the current paper attempted to propose a solution to assess the effects
of traffic volume on routes. For this purpose, an innovative method was presented in
which the GPS data were used to prepare a data source to display routes, nodes, their
relationship, and the volume of route traffic. The data source obtained for traffic volume
was converted into a distance matrix, indicating the difference between the routes and
the traffic volume. Using DL algorithms, the distance matrix was achieved for the future
state. LSTM and CNN predicted the motion states of moving objects in the future.
Both algorithms' loss and accuracy metrics were obtained with the time series along
the routes. As expected, the performance of the selected models was acceptable, and
the obtained results were in good agreement with each other. Finally, the correlation
coefficient test determined the distance matrix of the current and future states. It was
found that the number of changes in the volume of one route affected the other routes.
Notably, increasing or decreasing the volume of moving objects in one route caused
alterations in the other routes. In other words, changes on one route of the roadmap
were responsible for the traffic variation on other routes. The vital information to reduce
urban traffic congestion, time, cost, energy, etc., was achieved by analyzing the impact on
present and future routes based on the traffic volume. The presented model can compete
with the other state-of-art ones and even outperform them, while many attempts can be

made in the future to propose more accurate models. The measures that can improve the efficiency of the proposed method are as follows:

In the proposed method, the differences in the routes were examined in terms of road traffic volume. The differences in the routes must be assessed as a combination of features such as interfering with the destination type, weather conditions, etc.

Another limitation of the proposed method is that a single feature of the route node point was only used in this model to predict subsequent routes. More features, such as the type of location and the characteristics of the traveled routes, can be considered in the future.

References

1. Yi, Z., Liu, X.C., Markovic, N., Phillips, J.: Inferencing hourly traffic volume using data-driven machine learning and graph theory. Comput. Environ. Urban. Syst **85**, 101548 (2021)
2. Saha, R., Tariq, M.T., Hadi, M.: Deep learning approach for predictive analytics to support diversion during freeway incidents. Transp. Res. Rec **2674**(6), 480–492 (2020)
3. Georgiou, H., et al.: Moving objects analytics: survey on future location & trajectory prediction methods [Unpublished manuscript]. arXiv preprint arXiv:1807.04639, pp. University of Piraeus (2018)
4. Miller, J.: Dynamically computing fastest paths for intelligent transportation systems. IEEE. Intell. Transp. Syst. Mag **1**(1), 20–26 (2009)
5. Mikluščák, T., Gregor, M., Janota, A.: Using neural networks for route and destination prediction in intelligent transport systems. In: Mikulski, J. (ed.) International Conference on Transport Systems Telematics. TST 2012: Telematics in the Transport Environment, pp. 380–387, Springer, Heidelberg (2012)
6. Chakka, V.P., Everspaugh, A., Patel, J.M.: Indexing large trajectory data sets with SETI. In: *CIDR*, Ed., p. 76, Citeseer, Asilomar, CA, USA (2003)
7. Zheng, Y., Zhou, X.: Computing with Spatial Trajectories, Springer Science & Business Media, Heidelberg, Germany (2011)
8. Li, L., Wang, S., Wang, F.-Y.: An analysis of taxi driver's route choice behavior using the trace records. IEEE. Trans. Intell. Transp. Syst **5**(2), 576–582 (2018)
9. Costa, V., Fontes, T., Costa, P.M., Dias, T.G.: Prediction of journey destination in urban public transport. In: Pereira, F., Machado, P., Costa, E., Cardoso, A. (eds.) Portuguese Conference on Artificial Intelligence. EPIA 2015: Progress in Artificial Intelligence, pp. 169–180, Springer, Cham, (2015)
10. Endo, Y., Nishida, K., Toda, H., Sawada, H.: Predicting destinations from partial trajectories using recurrent neural network. In: Kim, J., Shim, K., Cao, L., Lee, J.-G., Lin, X., Moon, Y.-S. (eds.) Advances in Knowledge Discovery and Data Mining, pp. 160–172. Springer International Publishing, Cham, Switzerland (2017)
11. Toqué, F., Côme, E., El Mahrsi, M.K., Oukhellou, L.: Forecasting dynamic public transport Origin-Destination matrices with long-Short term Memory recurrent neural networks. In: 2016 IEEE 19th International Conference on Intelligent Transportation Systems (ITSC), pp. 1071–1076. IEEE, Rio de Janeiro, Brazil (2016)
12. Abbasifard, M.R., Naderi, H., Alamdari, O.I.: efficient indexing for past and current position of moving objects on road networks. IEEE. Trans. Intell. Transp. Syst **19**(9), 2789–2800 (2018)
13. Schneider, J.W., Borlund, P.: Matrix comparison, Part 1: motivation and important issues for measuring the resemblance between proximity measures or ordination results. J. Am. Soc. Inf. Sci. Technol **58**(11), 1586–1595 (2007)

14. Smouse, P.E., Long, J.C., Sokal, R.R.: Multiple regression and correlation extensions of the mantel test of matrix correspondence. Syst. Zool **35**(4), 627–632 (1986)
15. Yoo, W., Kim, T.-W.: Statistical trajectory-distance metric for nautical route clustering analysis using cross-track distance. J. Comput. Des. Eng **9**(2), 731–754 (2022)
16. Ait-Ali, A., Eliasson, J.: The value of additional data for public transport origin–destination matrix estimation. Public. Transp **14**(2), 419–439 (2022)
17. Yazdani, M., Mojtahedi, M., Loosemore, M.: Enhancing evacuation response to extreme weather disasters using public transportation systems: a novel simheuristic approach. J. Comput. Des. Eng **7**(2), 195–210 (2020)
18. Kuang, L., Hua, C., Wu, J., Yin, Y., Gao, H.: Traffic volume prediction based on multi-sources GPS trajectory data by temporal convolutional network. Mob. Netw. Appl **25**(4), 1405–1417 (2020)
19. Ashbrook, D., Starner, T.: Using GPS to learn significant locations and predict movement across multiple users. Pers. Ubiquitous. Comput **7**(5), 275–286 (2003)
20. Alvarez-Garcia, J.A., Ortega, J.A., Gonzalez-Abril, L., Velasco, F.: Trip destination prediction based on past GPS log using a Hidden Markov Model. Expert. Syst. Appl **37**(12), 8166–8171 (2010)
21. Panahandeh, G.: Driver route and destination prediction. In: 2017 IEEE Intelligent Vehicles Symposium (IV), pp. 895–900, IEEE, Los Angeles, CA, USA (2017)
22. Simmons, R., Browning, B., Zhang, Y., Sadekar, V.: Learning to predict driver route and destination intent. In: 2006 IEEE Intelligent Transportation Systems Conference, pp. 127–132, IEEE, Toronto, ON, Canada (2006)
23. Zong, F., Tian, Y., He, Y., Tang, J., Lv, J.: Trip destination prediction based on multi-day GPS data. Phys. A: Stat. Mech. Appl **515**, 258–269 (2019)
24. Kamble, S.J., Kounte, M.R.: On road intelligent vehicle path predication and clustering using machine learning approach. In: 2019 Third International conference on I-SMAC (IoT in Social, Mobile, Analytics and Cloud) (I-SMAC), pp. 501–505, IEEE, Palladam, India (2019)
25. Marmasse, N., Schmandt, C.: A user-centered location model. Pers. Ubiquitous. Comput **6**(5), 318–321 (2002)
26. Krumm, J.: A Markov model for driver turn prediction. In: Society of Automotive Engineers (SAE) 2008 World Congress, Published by SAE 2008 World Congress, Detroit, MI, USA (2008)
27. Froehlich, J., Krumm, J.: Route prediction from trip observations. In: Society of Automotive Engineers (SAE) 2008 World Congress, Published by SAE 2008 World Congress, Detroit, MI, USA (2008)
28. Liu, Q., Wu, S., Wang, L., Tan, T.: Predicting the next location: a recurrent model with spatial and temporal contexts. In: Thirtieth AAAI Conference on Artificial Intelligence, AAAI Publications, Phoenix, Arizona, USA (2016)
29. Laasonen, K.: Route prediction from cellular data. In: Proceedings of the workshop on context awareness for proactive systems CAPS 2005, pp. 147–157, Citeseer, Helsinki, Finland, 2005
30. Epperlein, J.P., Monteil, J., Liu, M., Gu, Y., Zhuk, S., Shorten, R.: Bayesian classifier for Route prediction with Markov chains. In: 2018 21st International Conference on Intelligent Transportation Systems (ITSC), pp. 677–682, IEEE, Maui, HI, USA (2018)
31. Vahedian, A., Zhou, X., Tong, L., Li, Y., Luo, J.: Forecasting gathering events through continuous destination prediction on big trajectory data. In: Proceedings of the 25th ACM SIGSPATIAL International Conference on Advances in Geographic Information Systems, pp. 1–10, Association for Computing Machinery, Redondo Beach, CA (2017)
32. de Sousa, R.S., Boukerche, A., Loureiro, A.A.F.: On the prediction of large-scale road-network constrained trajectories. Comput. Netw **206**, 108337 (2022)

33. De Brébisson, A., Simon, É., Auvolat, A., Vincent, P., Bengio, Y.: Artificial Neural Networks Applied to Taxi Destination Prediction [Unpublished manuscript], arXiv preprint arXiv:1508. 00021, pp. University of Montréal (2015)
34. Ke, J., Zheng, H., Yang, H., Chen, X.: Short-term forecasting of passenger demand under on-demand ride services: a spatiotemporal deep learning approach. Transp. Res. Part. C: Emerg. Technol **85**, 591–608 (2017)
35. Zhao, S., Zhao, P., Cui, Y.: A network centrality measure framework for analyzing urban traffic flow: a case study of Wuhan, China. Phys. A: Stat. Mech. Appl **478**, 143–157 (2017)
36. Choi, S., Yeo, H., Kim, J.: Network-wide vehicle trajectory prediction in urban traffic networks using deep learning. Transp. Res. Rec **2672**(45), 173–184 (2018)
37. Dai, G., Ma, C., Xu, X.: Short-term traffic flow prediction method for urban road sections based on space-time analysis and GRU. IEEE. Access **7**, 143025–143035 (2019)
38. Gao, S., Wang, Y., Gao, Y., Liu, Y.: Understanding urban traffic-flow characteristics: a rethinking of betweenness centrality. Environ. Plann. B. Plann. Des **40**(1), 135–153 (2013)
39. Huang, D., Deng, Z., Wan, S., Mi, B., Liu, Y.: Identification and prediction of urban traffic congestion via cyber-physical link optimization. IEEE. Access **6**, 63268–63278 (2018)
40. Yang, Y., Jia, L., Qin, Y., Han, S., Dong, H.: Understanding structure of urban traffic network based on spatial-temporal correlation analysis. Mod. Phys. Lett. B **31**(22), 1750230 (2017)
41. Kothai, G., et al.: A new hybrid deep learning algorithm for prediction of wide traffic congestion in smart cities. Wirel. Commun. Mob. Comput **2021**, 5583874 (2021)
42. Lohrasbinasab, I., Shahraki, A., Taherkordi, A., Delia Jurcut, A.: From statistical- to machine learning-based network traffic prediction. Trans. Emerg. Telecommun. Technol. **33**(4), e4394 (2022)
43. Ke, R., Li, W., Cui, Z., Wang, Y.: Two-stream multi-channel convolutional neural network for multi-lane traffic speed prediction considering traffic volume impact. Transp. Res. Rec **2674**(4), 459–470 (2020)
44. Yin, X., Wu, G., Wei, J., Shen, Y., Qi, H., Yin, B.: Deep learning on traffic prediction: methods, analysis, and future directions. IEEE Trans. Intell. Transp. Syst **23**(6), 4927–4943 (2022)
45. Tsai, C.-W., Hsia, C.-H., Yang, S.-J., Liu, S.-J., Fang, Z.-Y.: Optimizing hyperparameters of deep learning in predicting bus passengers based on simulated annealing. Appl. Soft Comput **88**, 106068 (2020)
46. Federgruen, A., Simchi-Levi, D.: Chapter 4 analysis of vehicle routing and inventory-routing problems. In: Handbooks in Operations Research and Management Science, Ed., pp. 297–373. Elsevier (1995)
47. O'Shea, K., Nash, R.: An Introduction to Convolutional Neural networks. [Unpublished manuscript]," arXiv preprint arXiv:1511.08458, pp. Department of Computer Science, Aberystwyth University (2015)
48. Shi, X., Chen, Z., Wang, H., Yeung, D.-Y., Wong, W.-K., Woo, W.-C.: Convolutional LSTM network: A machine learning approach for precipitation nowcasting. In: NIPS'15: Proceedings of the 28th International Conference on Neural Information Processing Systems, Ed., pp. 802–810. MIT Press, Montreal, Quebec, Canada (2015)
49. Baek, T., Lee, Y.-G.: Traffic control hand signal recognition using convolution and recurrent neural networks. J. Comput. Des. Eng **9**(2), 296–309 (2022)
50. Noh, S., An, K.: Reliable, robust, and comprehensive risk assessment framework for urban autonomous driving. J. Comput. Des. Eng **9**(5), 1680–1698 (2022)
51. Bogaerts, T., Masegosa, A.D., Angarita-Zapata, J.S., Onieva, E., Hellinckx, P.: A graph CNN-LSTM neural network for short and long-term traffic forecasting based on trajectory data. Transp. Res. Part. C: Emerg. Technol **112**, 62–77 (2020)
52. Zhao, J., et al.: Truck traffic speed prediction under non-recurrent congestion: based on optimized deep learning algorithms and GPS data. IEEE Access **7**, 9116–9127 (2019)

53. Ganapathy, J., Sureshkumar, T., Prasad, M.R., Dhamini, C.: Auto-encoder LSTM for learning dependency of traffic flow by sequencing spatial-temporal traffic flow rate: a speed-up technique for routing vehicles between origin and destination. In: 2022 International Conference on Innovative Trends in Information Technology (ICITIIT), Ed., pp. 1–6 (2022)
54. Zhang, Y., Lu, Z., Wang, J., Chen, L.: FCM-GCN-based upstream and downstream dependence model for air traffic flow networks. Knowl.-Based Syst. **260**, 110135 (2023)
55. Qi, H., Yao, Y., Zhao, X., Guo, J., Zhang, Y., Bi, C.: Applying an interpretable machine learning framework to the traffic safety order analysis of expressway exits based on aggregate driving behavior data. Physica A **597**, 127277 (2022)

Road Sign Classification Using Transfer Learning and Pre-trained CNN Models

Seyed Hossein Hosseini[1], Foad Ghaderi[1]([✉]) [iD], Behzad Moshiri[2] [iD],
and Mojtaba Norouzi[3]

[1] Human Computer Interaction Lab, Faculty of Electrical and Computer Engineering,
Tarbiat Modares University, Tehran, Iran
{h.seyedhossein,fghaderi}@modares.ac.ir
[2] School of Electrical and Computer Engineering, University of Tehran, Tehran, Iran
moshiri@ut.ac.ir
[3] School of ECE, University of Science and Technology, Tehran, Iran
mojtaba_norouzi77@elec.iust.ac.ir

Abstract. In this paper, we propose a transfer learning-based approach for road sign classification using pre-trained CNN models. We evaluate the performance of our fine-tuned VGG-16, VGG-19, ResNet50 and EfficientNetB0 models on the German Traffic Sign Recognition Benchmark (GTSRB) test dataset. Our work makes several contributions, including the utilization of transfer learning with pre-trained CNN models, the integration of augmentation techniques, and a comprehensive evaluation of multiple state-of-the-art architectures on the road sign classification task. Our experimental results show that the proposed approach achieves an average accuracy of 99.2% on the GTSRB test dataset, outperforming existing state-of-the-art approaches. The results confirm that transfer learning and pre-trained models can significantly improve the accuracy of road sign classification, even with a relatively small dataset.

Our study contributes to the field of intelligent transportation systems by providing an efficient and accurate method for road sign classification. We believe that our approach can be easily extended to other computer vision tasks and applied to real-world scenarios.

Keywords: Road sign classification · Transfer learning · Convolutional neural networks · Autonomous Vehicles

1 Introduction

Road signs play a crucial role in ensuring road safety by providing essential information to drivers about road conditions, hazards, and regulations. However, the accurate and timely recognition of road signs can be challenging, due to variations in lighting, weather conditions, and occlusions or in complex traffic scenarios [1]. Therefore, road sign classification is an important problem in computer vision and it has a wide range of practical applications, including autonomous driving, traffic management, and road safety. Automatic road sign detection and recognition can assist drivers in making more informed

M. Ghatee and S. M. Hashemi (Eds.): ICAISV 2023, CCIS 1883, pp. 39–52, 2023.
https://doi.org/10.1007/978-3-031-43763-2_3

decisions, reduce accidents, and improve overall traffic flow. In recent years, convolutional neural networks (CNNs) have emerged as a powerful tool for image classification tasks, including road sign classification. CNNs are well-suited for this task because they can automatically learn features from the input images and capture complex patterns and structures [15].

Transfer learning is a popular technique in deep learning that utilizes pre-trained CNN models on large-scale datasets to improve the performance of a smaller dataset with similar features.

In this paper, we investigate the use of transfer learning and pre-trained CNN models for road sign classification. Due to memory limitations, we use the German Traffic Sign Recognition Benchmark (GTSRB) test dataset, which consists of more than 12,000 images of 43 different road sign classes, to evaluate the performance of several pre-trained CNN models. We also fine-tuned these models on the GTSRB dataset and achieved an accuracy of up to 99%.

Our experimental results demonstrate that pre-trained models with fine-tuning and transfer learning outperform models trained from scratch and traditional machine learning approaches, achieving higher accuracy with fewer training epochs, while effectively leveraging pre-existing knowledge to enhance the accuracy and efficiency of road sign recognition systems [16].

Our work makes several contributions, including the utilization of transfer learning with pre-trained CNN models, the integration of augmentation techniques, and a comprehensive evaluation of multiple state-of-the-art architectures on the road sign classification task.

This paper is organized as follows: in Sect. 2, we provide a review of the related work on road sign classification using transfer learning and pre-trained models. Section 3 describes the GTSRB dataset and the preprocessing techniques employed and presents the details of the pre-trained models used in our experiments and the fine-tuning process. In Sect. 4, we present the experimental results and analyze the performance of our approach. Finally, we conclude the paper in Sect. 5 with a summary of our contributions, limitations, and future research directions.

2   Related Works

Road sign recognition is an active area of research, with many approaches proposed in recent years. Deep learning, specifically CNNs, has emerged as a powerful tool for road sign classification due to its ability to learn features directly from the images. In addition, transfer learning, which involves leveraging pre-trained CNN models on large-scale datasets, has been shown to be effective in improving the performance of CNNs on smaller datasets, such as road sign datasets [14].

In recent years, a growing body of research has focused on developing CNN-based approaches for road sign recognition, with the goal of improving the accuracy and speed of existing systems [13]. In this section, we provide a brief review of the most relevant and recent studies that have investigated road sign classification using CNNs, with a focus on approaches that employ transfer learning and pre-trained models.

Transfer learning has been used in several studies to improve the accuracy and efficiency of road sign recognition systems.

In [2], the authors proposed a CNN-based approach for traffic sign recognition using the Chinese Traffic Sign Recognition dataset. They utilized a pre-trained Inception-v3 model and fine-tuned it on the dataset. Their approach achieved an accuracy of 97.78%, outperforming existing approaches based on hand-crafted features.

In [3], the authors proposed a deep learning approach based on residual networks (ResNets) for road sign classification using the GTSRB dataset. They demonstrated that ResNet models outperform traditional machine learning approaches and achieved state-of-the-art accuracy on the GTSRB dataset.

In [4], the authors proposed an optimized CNN architecture for classification. Using this architecture on the GTSRB dataset, they were able to achieve the accuracy of 98.9%.

Qian et al. proposed CNNs based on Max Pooling Positions (MPPs) for testing on GTSRB dataset. They achieved an accuracy of 98.86% [5].

In addition to transfer learning, other techniques have been proposed to improve the performance of CNNs in road sign recognition. For example, Zhou et al. proposed a region-based CNN approach that is called PFANet. This approach achieved an accuracy of 97.21% on the GTSRB dataset [6].

Several studies have investigated the impact of different CNN architectures on road sign classification. Alawaji et al. compared the performance of several CNN architectures and found that InceptionResNetV2 achieved the highest accuracy of 98.84% [7]. Additionally, the authors of [8] proposed an European dataset and tested several CNN models on it. They explored the performance of various CNN architectures, including the 8-layer CNN model, which provided an accuracy of 98.52%.

In a related study conducted by authors [11], a CNN model was proposed for the road sign classification task, employing GTSRB test dataset. The experimental results revealed a notable accuracy of 98.4% attained by their model.

Overall, these studies demonstrate the effectiveness of CNN-based approaches for road sign classification and the potential for incorporating transfer learning and other advanced techniques to improve the accuracy and speed of existing systems.

Existing approaches suffer from computational complexity, and generalization. To address these issues, our method employs transfer learning with pre-trained CNN models for efficient feature extraction. We also utilize augmentation techniques to enhance data diversity. Fine-tuning enables the adaptation of the pre-trained models for road sign classification. Our method aims to overcome these limitations by achieving higher accuracy, improved efficiency, and enhanced generalization.

In order to validate the superiority of our proposed method over traditional deep learning approaches, we conducted a series of experiments comparing their performance [12]. Specifically, we trained and evaluated several traditional deep learning models, including simple CNN architectures, using the same dataset and evaluation criteria as our proposed method. For example, we trained and subsequently evaluated a CNN model on our dataset, resulting in an achieved accuracy of approximately 96.7%. The results of these experiments provide a comprehensive basis for the comparison and allow us to assess the effectiveness and advancements offered by our proposed approach.

Our research is built upon this prior work by investigating the performance of several pre-trained CNN models with fine-tuning on the GTSRB dataset.

3 Methodology

3.1 Data Preprocessing

The German Traffic Sign Recognition Benchmark (GTSRB) test dataset was used for this study, which consists of more than 12,000 images of 43 different road sign classes. The images have varying resolutions and aspect ratios and were captured under different lighting and weather conditions.

The dataset was preprocessed using several techniques to improve the performance of the CNN models and prevent overfitting. The dataset was split into 80% training data and 20% testing data. To prepare the data for training, we performed data augmentation using several techniques. Firstly, we randomly rotated each image between -15 and 15 degrees to increase the variety of angles that the model would be exposed to during training. Additionally, we randomly cropped each image by a factor of 0.1 to 0.2, allowing the model to learn recognizing signs even when they are partially obscured. We also randomly flipped each image horizontally to further increase the diversity of the training data. We also used the zooming technique. Zooming with a scale of 10% is an image augmentation technique that randomly zooms the input image by a factor of 0.9 to 1.0. This technique adds more variety to the training data by artificially creating images that are closer to or farther away from the camera than the original image. This technique is particularly useful for improving the model's ability to recognize objects at different distances. Additionally, we have used the image shift technique. Shift augmentation is another technique that can be used to create additional training examples. It randomly shifts the input image in the width and height dimensions by up to 10%. This technique is used to create new images that have different compositions and backgrounds, which can improve the model's ability to recognize objects under different lighting conditions or against different backgrounds. Furthermore, we balanced the dataset to ensure that each class had an equal number of images during training. This was done by undersampling the overrepresented classes and oversampling the underrepresented classes until each class had the same number of images. This technique helps to prevent bias towards the majority classes during training and improves the performance of the model on the minority classes. Finally, approximately 293 images are placed in each class equally. After preprocessing, each image was resized to a specific size of each models input size to ensure that the images were of consistent size.

In summary, the GTSRB dataset was preprocessed using data augmentation techniques such as random rotation, cropping and flipping as well as balancing the dataset to ensure each class had an equal number of images. These preprocessing steps were essential to increase the diversity of the training data and to avoid overfitting during training.

3.2 Transfer Learning

Transfer learning has emerged as a powerful technique in the field of deep learning, allowing us to leverage knowledge learned from one task to another related task. In the context of computer vision, transfer learning involves using pre-trained convolutional neural network (CNN) models that have been trained on large-scale image datasets such as ImageNet to solve new image classification problems with limited training data. By fine-tuning the pre-trained model on the new dataset, we can improve the accuracy and reduce the training time compared to training from scratch. The core idea behind transfer learning is that the low-level features learned by the early layers of a CNN, such as edge detection and texture recognition, are generic and transferable to other tasks. Therefore, we can use these pre-trained weights as a starting point and fine-tune the higher layers of the network for the new task. This process of freezing the early layers and training the later layers is called fine-tuning. There are several benefits of using transfer learning for image classification tasks. First, pre-trained models are trained on massive amounts of data, which enables them to learn high-level features that can be used for various tasks. Second, by using transfer learning, we can avoid the computational cost and time-consuming process of training a model from scratch. Finally, transfer learning can improve the accuracy of the model, especially when we have limited training data [14]. In this paper, we utilized transfer learning by fine-tuning several pre-trained CNN models on the GTSRB test dataset. By leveraging the knowledge learned from these pre-trained models, we achieved high accuracy in road sign classification even with a relatively small dataset. Our results demonstrate the effectiveness of transfer learning in improving the accuracy of image classification tasks [9].

3.3 Method Details

In this section, we describe the methodology used for our road sign classification task using transfer learning and pre-trained CNN models, that were originally trained on ImageNet, on the GTSRB test dataset. Our approach involves using pre-trained CNN models which were trained on large-scale datasets such as ImageNet, as feature extractors, followed by a fully connected layer for classification. These models were chosen because they have shown promising results in image classification tasks and they are widely used in the literature for road sign recognition.

We first load the pre-trained CNN models and remove the last few layers of the network, including the fully connected layers, which are responsible for classifying the original task of the network. We then freeze the weights of the remaining layers and use them as feature extractors for our road sign classification task. Next, we add a new fully connected layer to the pre-trained model, which takes the extracted features as input and outputs the probability distribution over the 43 road sign classes. We train this new layer using the extracted features and the corresponding labels of the training set.

We used the Adam optimizer with a learning rate of $1e^{-5}$ and a batch size of 32 for training our models. We trained each model for 15 epochs and monitored the validation accuracy to avoid overfitting. In order to avoid overfitting, we also used early stopping with a patience of 3 epochs to stop training if the validation accuracy did not improve for 3 consecutive epochs.

For the fine-tuning process, we conducted a total of 20 runs on each model to ensure the convergence and stability of the models. Each run involved fine-tuning the pre-trained CNN models using our road sign dataset. By performing multiple runs, we aimed to capture the potential variations in model performance due to different initializations and training processes. Furthermore, to select the best configuration, we ran the fine-tuned models three additional times and evaluated their performance. This approach allowed us to identify the most reliable and consistent results, considering the potential variability in the fine-tuning process. The combination of multiple runs and the selection of the best configuration helped us to ensure the reliability and robustness of the reported results.

We evaluate the performance of our models using various metrics, including accuracy, precision, recall, and F1-score for each model. We also performed a comparative analysis of the models to determine which model achieved the highest performance. In Sect. 4, the results of the models are presented in detail.

The overall process of our research involves several steps, including data preprocessing, training and fine-tuning pre-trained CNN models, and evaluating model performance. Figure 1 provides a visual overview of our methodology.

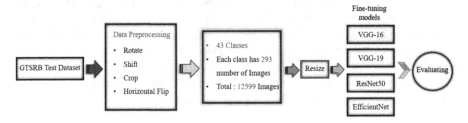

Fig. 1. Methodology Overview

4   Results and Analysis

4.1   Performance and Metrics

In this section, we present the results of our road sign classification task using transfer learning and pre-trained CNN models.

We fine-tuned four pre-trained models, including VGG-16, VGG-19, ResNet50 and EfficientNetB0 on the GTSRB dataset.

Table 1 shows the accuracy, precision, recall, and F1 score for each model. The results show that all models achieved high accuracy, with the pre-trained VGG-16 model achieving the highest accuracy of 99.21 followed closely by VGG-19 and RseNet50 with accuracies of 98.98% and 98.55%.

Table 1. Performance Metrics of Road Sign Classification Models

Model	Accuracy	Precision	F1-Score
VGG-16	99.21	99.40	99.11
VGG-19	98.98	99.12	99.02
ResNet50	98.85	98.49	98.68
EfficientNetB0	98.77	99.26	99.09

Overall, our results demonstrate that transfer learning and pre-trained CNN models can be effectively used for road sign classification tasks, achieving high accuracy even with a relatively small dataset. The VGG-16 model performed the best in our experiments, but the other pre-trained models also achieved excellent performance.

4.2 Class Metrics

To further analyze the performance of each model, we evaluated each model on the test set and computed the precision, recall, F1 score, and support for each class.

All models achieved high precision, recall, and F1 score across all classes. The models performed particularly well on classes with high support, indicating that the models are able to generalize well to common classes. However, some models performed slightly better than others on specific classes, indicating that certain models may be better suited for certain types of road signs. In summary, the VGG-16 and VGG-19 models achieved the highest F1-score for most of the classes.

4.3 Accuracy and Loss Plots

The accuracy plot shows how well the model is able to correctly classify the images in the test dataset over time, as the number of epochs increases. As the model trains, it becomes better and better at recognizing the patterns in the data and making accurate predictions. This is reflected in the steadily increasing accuracy over time. The loss plot, on the other hand, shows how well the model is able to minimize its error during training. The loss function is a measure of how well the predicted values match the actual values, so a lower loss indicates that the model is doing a better job of fitting the data. As the number of epochs increases, the model learns to better generalize the patterns in the data, resulting in a steady decrease in the loss [10].

As shown in Fig. 2 the plot shows the accuracy of the ResNet50 model during the training and validation phases. The training accuracy is calculated by measuring the accuracy of the model on the training dataset, while the validation accuracy is calculated on a separate validation dataset. As we can see from the plot, the accuracy of the models steadily increases during the training phase and eventually plateaus. This indicates that the model has learned to classify the images accurately.

As shown in Fig. 3 the plot shows the loss of the ResNet50 model during the training and validation phases. The loss represents the difference between the predicted output and the actual output. As we can see from the plot, the loss decreases gradually during the training phase, which means the model is improving in its ability to classify images.

Accuracy and loss plots for the other models are included in the paper to provide a comprehensive analysis of their performance. These plots are as follows (Figs. 5, 6, 7, 8 and 9):

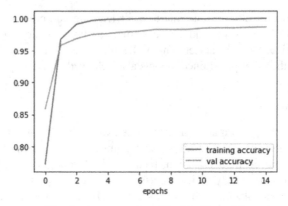

Fig. 2. Accuracy plot of ResNet50 model

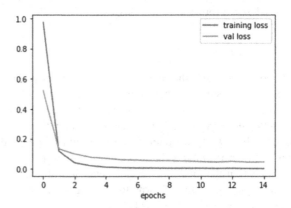

Fig. 3. Loss plot of ResNet50 model

4.4 Confusion Matrix

The confusion matrix can be used to calculate various performance metrics of a classification model, including precision, recall, F1 score, and accuracy.

Confusion matrices can provide important insights into the strengths and weaknesses of a classification model. By analyzing the confusion matrix, one can identify which classes are frequently confused with each other and adjust the model accordingly. It can also help identify where the model is making mistakes and suggest ways to improve the performance.

Figure 4 shows the confusion matrix for the VGG-19 model on the test set. The rows indicate the real labels of the classes, whereas the columns show the predicted labels of the classes. Each cell in the table represents the number of samples predicted for a given class. Based on the confusion matrix, it can be seen that the VGG-19 model has high accuracy for some classes such as "bicycle crossing" and "wild animals crossing", with accuracies of 100% and 99%, respectively. However, it also has relatively lower accuracy for some other classes such as "speed limit 100 km/h" and "No passing", with

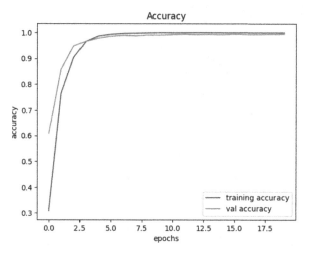

Fig. 4. Accuracy plot of VGG-16 model

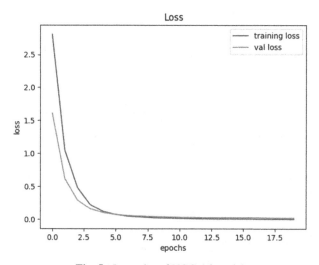

Fig. 5. Loss plot of VGG-16 model

accuracies of 0.95% and 98%, respectively. It is also observed that some classes are more commonly confused with other classes, such as "speed limit 80 km/h" is being confused with "speed limit 100 km/h" and "speed limit 50 km/h" is being confused with "speed limit 60 km/h". This information can be used to improve the model by incorporating additional training data or adjusting the model architecture. Overall, the confusion matrix for the VGG-19 model demonstrates its ability to accurately classify the traffic signs, while also revealing areas where it can be improved (Fig. 10).

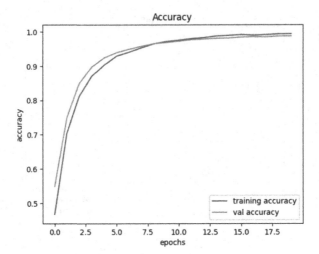

Fig. 6. Accuracy plot of VGG-19 model

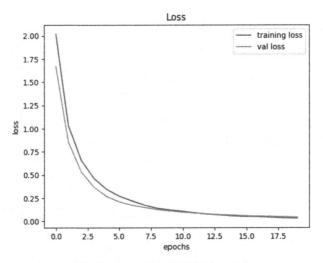

Fig. 7. Loss plot of VGG-19 model

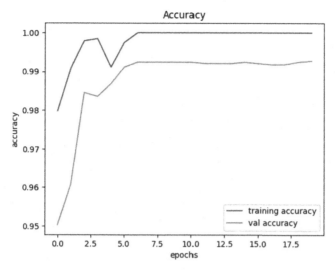

Fig. 8. Accuracy plot of EfficientNetB0 model

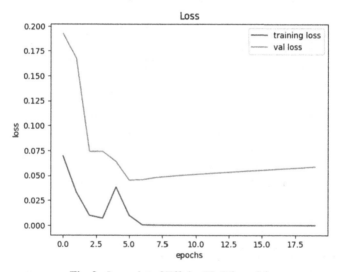

Fig. 9. Loss plot of EfficientNetB0 model.

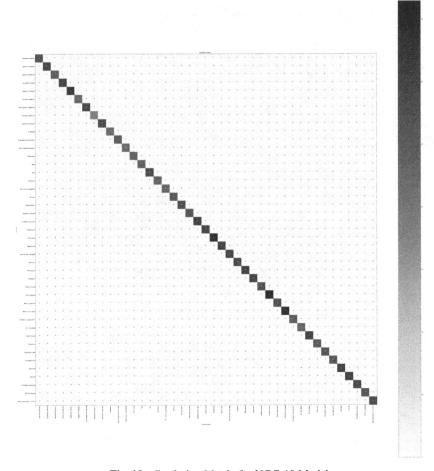

Fig. 10. Confusion Matrix for VGG-19 Model

5 Conclusion

Our results demonstrate that transfer learning and pre-trained models can significantly improve the performance of road sign classification tasks. Furthermore, we showed that pre-trained models can be fine-tuned on a small dataset, such as the GTSRB test dataset, with excellent results. We also demonstrated the importance of data preprocessing techniques such as data augmentation and dataset balancing, which can help improve the accuracy and robustness of the models.

The innovation of this paper lies in the successful application of transfer learning and pre-trained CNN models to the task of road sign classification using the GTSRB dataset. By leveraging the powerful features learned from pre-trained models, the proposed method achieves high accuracy of around 99%, which is a significant improvement compared to existing approaches. Additionally, we used good data augmentation techniques to improve the robustness and generalization ability of our models so this makes these models able to perform well even in challenging conditions.

This study demonstrates that by using transfer learning and pre-trained CNN models with fine tuning and image augmentation techniques and balancing the dataset, it is possible to achieve state-of-the-art performance in road sign classification.

Our work has important implications for real-world applications such as self-driving cars, traffic management systems, driver assistance systems, where accurate road sign classification is crucial. The high accuracy of our models suggests that transfer learning and pre-trained models can be effective in such applications.

6   Future Works

Although our proposed approach achieved high accuracy in classifying road signs, there is still room for improvement.

- We propose exploring image fusion techniques and decision fusion principles to combine the results of multiple models and achieve better accuracy. These techniques have shown promising results in other computer vision tasks and can potentially boost the accuracy of our road sign classification system.
- We suggest investigating the use of more advanced pre-processing techniques, such as noise reduction, feature selection, and outlier detection, to further improve the quality of the input data and reduce the noise and variability in the signals.
- We can evaluate the proposed method in real-world applications. The ultimate goal of road sign classification is to enhance the safety and efficiency of driving. Therefore, future work could involve testing the proposed method in real-world driving scenarios to evaluate its effectiveness in improving driving performance and safety.

References

1. Wali, S.B., et al.: Vision-based traffic sign detection and recognition systems: current trends and challenges. Sensors **19**, 2093 (2019)
2. Kim, C.-I., Park, J., Park, Y., Jung, W., Lim, Y.-S.: Deep learning-based real-time traffic sign recognition system for urban environments. Infrastructures **8**, 20 (2023). https://doi.org/10.3390/infrastructures8020020
3. Rawat, W., Wang, Z.: Deep convolutional neural networks for image classification: a comprehensive review. Neural Comput. **29**, 2352–2449 (2017)
4. Wong, A., Shafiee, M.J., St. Jules, M.: MicronNet: a highly compact deep convolutional neural network architecture for real-time embedded traffic sign classification. In: IEEE Access, vol. 6, pp. 59803–59810 (2018). Doi: https://doi.org/10.1109/ACCESS.2018.2873948
5. Qian, R., Yue, Y., Coenen, F., Zhang, B.: Traffic sign recognition with convolutional neural network based on max pooling positions. In: 2016 12th International Conference on Natural Computation, Fuzzy Systems and Knowledge Discovery (ICNC-FSKD), Changsha, China, 2016, pp. 578–582 (2016). https://doi.org/10.1109/FSKD.2016.7603237
6. Zhou, K., Zhan, Y., Fu, D.: Learning region-based attention network for traffic sign recognition. Sensors **21**(3), 686 (2021)

7. Alawaji, K., Hedjar, R.: Comparison study of traffic signs recognition using deep learning architectures. In: 2022 13th International Conference on Information and Communication Systems (ICICS), Irbid, Jordan, pp. 442–447 (2022). https://doi.org/10.1109/ICICS55353.2022.9811216

8. Gámez Serna, C., Ruichek, Y.: Classification of traffic signs: the European dataset. IEEE Access **6**, 78136–78148 (2018). https://doi.org/10.1109/ACCESS.2018.2884826

9. Bouaafia, S., Messaoud, S., Maraoui, A., Ammari, A.C., Khriji, L., Machhout, M.: Deep pre-trained models for computer vision applications: traffic sign recognition. In: 2021 18th International Multi-Conference on Systems, Signals & Devices (SSD), Monastir, Tunisia, 2021, pp. 23–28 (2021). https://doi.org/10.1109/SSD52085.2021.9429420

10. Chollet, F.: Xception: deep learning with depthwise separable convolutions. In: 2017 IEEE Conference on Computer Vision and Pattern Recognition (CVPR), Honolulu, HI, USA, 2017, pp. 1800–1807 (2017). https://doi.org/10.1109/CVPR.2017.195

11. Vincent, M., Vidya, K.R., Mathew, S.P.: Traffic sign classification using deep neural network. In: 2020 IEEE conference on Recent Advances in Intelligent Computational Systems (RAICS), Trivandrum, India (2020)

12. Bouti, A., Mahraz, M.A., Riffi, J., Tairi, H.: A robust system for road sign detection and classification using LeNet architecture based on convolutional neural network. Soft. Comput. **24**(9), 6721–6733 (2020)

13. Swaminathan, V., Arora, S., Bansal, R., Rajalakshmi, R.: Autonomous driving system with road sign recognition using convolutional neural networks. In: 2019 International Conference on Computational Intelligence in Data Science (ICCIDS), pp. 1–4. IEEE, February 2019

14. Bouaafia, S., Messaoud, S., Maraoui, A., Ammari, A. C., Khriji, L., Machhout, M.: Deep pre-trained models for computer vision applications: traffic sign recognition. In 2021 18th International Multi-Conference on Systems, Signals & Devices (SSD), pp. 23–28. IEEE, March 2021

15. Gadri, S., Adouane, N.E.: Efficient traffic signs recognition based on cnn model for self-driving cars. In: Intelligent Computing & Optimization: Proceedings of the 4th International Conference on Intelligent Computing and Optimization 2021 (ICO2021) 3, pp. 45–54. Springer International Publishing (2022)

16. Haque, W.A., Arefin, S., Shihavuddin, A.S.M., Hasan, M.A.: DeepThin: a novel lightweight CNN architecture for traffic sign recognition without GPU requirements. Expert Syst. Appl. **168**, 114481 (2021)

Improving Safe Driving with Diabetic Retinopathy Detection

Niusha Sangsefidi[1]([⊠]) and Saeed Sharifian[2]

[1] Department of Management, Science and Technology, Amirkabir University, Tehran, Iran
niusha.sangsefidi@aut.ac.ir
[2] Department of Electrical Engineering, Amirkabir University, Tehran, Iran
sharifian_s@aut.ac.ir

Abstract. Driving requires significant cognitive and physical capabilities. Driving performance may be impaired by various complications of diabetes, such as vision impairment. Diabetic retinopathy is the most prevalent eye disease caused by diabetes. It can result in cloudy vision or even blindness if not diagnosed early enough. However, detecting the disease in its early stages can be challenging because of the absence of noticeable symptoms. Although there are existing models for diabetic retinopathy, they are not capable of detecting all stages of the disease. In this paper, we propose a ResNet-50 deep neural network for feature extraction from retinal images and a SVM machine learning classifier to determine the degree of opacity for drivers with diabetes. The proposed method showed 79% accuracy in the APTOS dataset. We suggest a promising application for this model alongside smart glasses technology to ensure safe driving among diabetic retinopathy patients. This innovative approach would provide real-time updates about an individual's condition and enable them to take actions to maintain their and others' safety while driving.

Keywords: Diabetic Retinopathy · ResNet-50 · Transfer learning · SVM

1 Introduction

Driving is a crucial aspect of transportation for millions of people worldwide. Safe driving requires a combination of complex psychomotor skills, rapid information processing, vigilance, and sound judgment [1, 2]. Although driving is typically considered a light physical activity, studies conducted using driving simulators have shown that it also incurs significant metabolic demands, particularly on glucose consumption mainly in the brain [3].

Diabetes complications can threaten driving performance. It is widely acknowledged that diabetes contributes to many driving mishaps and accidents on the road [1, 2]. Safe driving performance depends heavily on clear vision, and any impairment can drastically affect drivers' abilities. Poor eyesight puts diabetic drivers at risk of car accidents [4]. Therefore, it is crucial for diabetes drivers to maintain a precise understanding of their disease level to ensure safe driving. But the main problem is that, most times, diabetes

© The Author(s), under exclusive license to Springer Nature Switzerland AG 2023
M. Ghatee and S. M. Hashemi (Eds.): ICAISV 2023, CCIS 1883, pp. 53–61, 2023.
https://doi.org/10.1007/978-3-031-43763-2_4

is asymptomatic [5]. Diabetic retinopathy, the prevalent eye disease among people with diabetes, is a primary contributor to vision loss in American adults [6]. The retina's blood vessels are damaged by high blood sugar levels, leading to two common types of retinopathy: non-proliferative (NPDR) and proliferative (PDR). NPDR is the early stage, where the retinal blood vessels weaken and leak fluid or blood. As the condition progresses, the blood vessels may become blocked, reducing oxygen supply to the retina. NPDR is classified into three levels: Mild, Moderate, and Severe. PDR is a more advanced stage, where new blood vessels grow in the retina, causing more bleeding and vision loss [4, 6, 7]. Diabetic Retinopathy levels are shown in Table 1. It is essential that diabetic patients undergo regular eye examinations to reduce their risk of retinopathy [8, 9]. In addition, the process also demands extensive knowledge and expensive machinery, which is unfortunately absent in underprivileged regions. Some examples of images that represent DR severity levels are shown in Fig. 1. The challenges associated with traditional grading of Diabetic Retinopathy (DR) can be resolved by an automated DR grading system. Given the high expertise and valuable equipment required for accurate DR analysis, the development of computer-aided diagnosis systems has become increasingly critical to shorten the time taken to diagnose patients. Deep learning advancements have opened up new possibilities in medical diagnosis, especially with the ever-increasing cases of diabetes worldwide [8, 10]. With its multiple layers of computational neural networks and impressive ability to analyze vast amounts of visual data, deep learning algorithms have become a leading approach in many fields especially in computer vision [8]. In this paper we propose a ResNet-50 convolutional neural network for feature extraction from retinal images and a SVM machine learning classifier to determine the degree of opacity for diabetes drivers. Our proposed approach has the potential to streamline screening and diagnosis of diabetic retinopathy, ultimately leading to earlier detection and more effective treatment.

Table 1. DR Grading [8, 11]

DR classes	Findings
class 0: No DR	Normal condition
class 1: Mild NPDR	Small amount of Microaneurysms
class 2: Moderate NPDR	considerable amount of aneurysms
class 3: Severe NPDR	When-some venous beading in two or more quadrants of retina image among over 20 intraretinal hemorrhages
class 4: PDR	One or combination of the following effects: vitreous hemorrhages, neovascularization and preretinal hemorrhages

Advancement in technology has provided various innovative solutions towards enhancing road safety. One such solution is the use of smart glasses with retinal imaging capabilities. With this method, it is possible to monitor the driver's vision status while driving by taking pictures of their retina and identifying eye diseases such as diabetic retinopathy. This information can then be used to alert and warn the driver about his/her condition. Thus, this approach is a critical step towards promoting driver safety and reducing accidents caused by impaired vision.

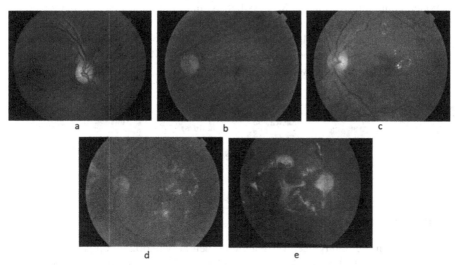

Fig. 1. Sample images illustrating the different severity levels of DR. a) No DR, b) Mild NPDR, c) Moderate NPDR, d) Severe NPDR, e) PDR [12].

2 Literature Review

In 2016, Doshi et al. proposed a method for automatic detection and classification of diabetic retinopathy using Deep Convolutional Neural Networks on the EyePACs dataset [9]. They developed three CNN models with QWK score of 0.3066, 0.35 and 0.386. The best score of 0.3996 is obtained by the ensemble of these three models.

In 2021, Yaqoob et al. used ResNet-50 architecture along with Random Forest as a classifier for automated detection of diabetic retinopathy on the Messidor-2 and Eye-PACS datasets [10]. They used dropout to reduce overfitting. Their approach achieved an accuracy of 96% on the Messidor-2 dataset with two classes and an accuracy of 75.09% on the EyePACS dataset with five classes.

In 2022, Chilukoti et al. used the resized version of the diabetic retinopathy Kaggle competition dataset for diabetic retinopathy detection [6]. They developed the Effcient-Net_b3_60 model, which detects all the five stages of diabetic retinopathy with a QWK score of 0.85.

In 2023, Alwakid et al. proposed an algorithm based on Inception-V3 that could detect diabetic retinopathy using the APTOS dataset [13]. Their suggested method presented two scenarios: case 1 with image enhancement using CLAHE and ESRGAN techniques, and case 2 without image enhancement. Augmentation techniques were used to generate a balanced dataset. Their model achieved an accuracy of 98.7% for case 1 and 80.87% for case 2.

3   Proposed Method

The proposed method relies on transfer learning and utilizes deep features from a ResNet-50 model, combined with an SVM classifier. At first, the input image undergoes pre-processing technique such as resizing and normalization that aim to reduce computational costs and improve accuracy. Then a ResNet model is used to extract significant features. Finally, the high-level extracted features are fed into a SVM classifier, which generates the class label. The proposed workflow is shown in Fig. 2.

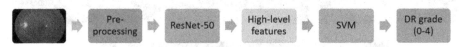

Fig. 2. Proposed workflow

3.1   ResNet Architecture

ResNet-50 is an incredibly deep convolutional neural network, comprising 50 layers. As part of the Residual Networks family, ResNet has become a prominent neural network used as a basis in many computer vision tasks. The ResNet architecture improves gradient flow by introducing residual connections that carry the original input to deeper layers and alleviates the vanishing gradient problem faced by deep neural networks. This widespread adoption attests to ResNet's effectiveness and its remarkable contribution to deep neural networks development today [14]. ResNet architecture is shown in Fig. 3.

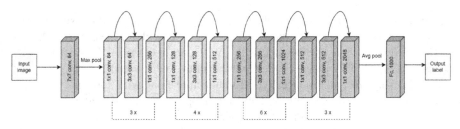

Fig. 3. ResNet Architecture

3.2   Baseline Model Description

We used transfer learning to reduce the training time of the model. We fine-tuned the pre-trained ResNet model by adding a few additional layers on top of it to adapt it to the new task. First, the ResNet base model generates features of size $4 \times 4 \times 2048$. The average pooling and flattened layers are used to extract 2048 features from these features. Next, the dropout layer (dropout rate $= 0.2$) and the ReLU activation function are applied. Finally, a fully connected layer with five outputs and a Softmax activation function is applied to determine probabilities associated with each retinopathy class. After model training, we regenerate a new network by providing the base model as input and the flattened layer as output. After that, we utilize this network to extract features from the training data. Finally, the extracted features are passed to SVM classifier to make a conclusive diagnosis of the image. The SVM algorithm is specifically designed for handling nonlinear data and is known for its ability to effectively deal with complex feature spaces. Our proposed architecture is shown in Fig. 4.

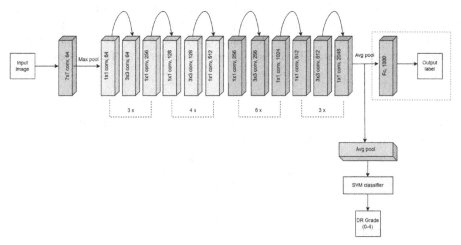

Fig. 4. Proposed Architecture

3.3 Proposed Model Improvement

We used the Adam optimizer which combines the advantages of RmsProp and momentum. This optimization algorithm tunes learning rate according to weight gradients while also allowing the model to overcome local minima. Moreover, we use early stopping technique to prevent overfitting. Using this technique, the model can achieve better generalization and capture the underlying patterns in the data, rather than simply fitting to the noise in the training set. In addition, we employ the strategy of reducing the learning rate, a widely used technique in guiding machine learning models towards achieving optimal performance. By reducing the learning rate during the training process, we allow the optimization algorithm to take smaller steps towards the optimum. This enables the model to more accurately traverse the optimization landscape and avoid local minima.

4   Experiments and Results

This study utilizes the APTOS [12] dataset to evaluate the proposed method. In this section, we provide a detailed overview of this dataset and performance metrics.

4.1 Dataset

The APTOS [12] dataset contains 3662 retinal images. The images were captured under varying conditions, including differing levels of illumination, focus, and noise. However, the dataset is well-curated and labeled, enabling robust assessment and comparison of the performance of different machine learning algorithms on this task. As mentioned in Table 1, we have five classes. Table 2 shows the distribution of APTOS classes.

4.2 Evaluation Metric

Accuracy is considered as a key metric for the multi-class classification performance of the models. The following equation is used to compute accuracy [15]:

$$Accuracy = \frac{\text{TP} + \text{TN}}{\text{TP} + \text{TN} + \text{FP} + \text{FN}} \tag{1}$$

4.3 Experimental Results

In this article, we explored various classifiers, examined their performance, and compared their results.

Environment. To conduct the experimentation, Google Colab was utilized, as it provides free GPU in the cloud. It is well-suited to the highly computational nature of the experiments. Furthermore, Google Colab simplifies working with data, as it is connected to Google Drive.

Results. The model presented in this study achieved an accuracy of 79% in grading diabetic retinopathy, which is considered a promising result relative to other approaches, given the fact that the images were taken under different conditions. Thus, the model's performance can be deemed acceptable. This approach combines both SVM and the ResNet-50 architecture to improve the model's performance. Table 3 displays the outcomes obtained from various classifiers. The confusion matrix and loss and accuracy diagrams are shown in Figs. 5 and 6, respectively.

Table 2. Distribution of classes in the APTOS dataset

Class	Count
No-DR	1805
Mild-NPDR	370
Moderate-NPDR	999
Severe-NPDR	193
PDR	295

Table 3. Comparison of proposed approach with different classifiers

Classifier	Accuracy
SVM with linear kernel	79.01%
SVM with RBF kernel	69.75%
Decision Tree	72.75%
Random Forest	76.83%
Gaussian Naive Bayes	56.13%
KNN	76.02%
Ada Boost	73.29%
MLP	78.20%

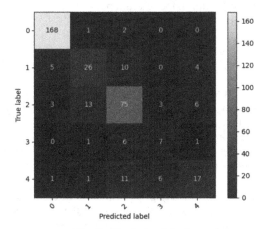

Fig. 5. Confusion Matrix

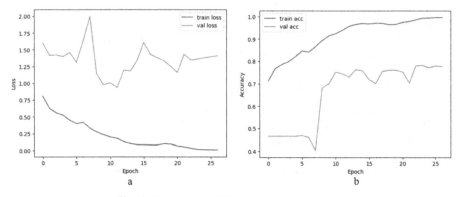

Fig. 6. Train and validation a) Loss, b) Accuracy

5 Conclusions

In this paper, we introduce a novel approach to diagnosing and grading diabetic retinopathy by proposing a ResNet-50-based convolutional neural network model. The proposed model achieved 79% accuracy on the APTOS dataset. Utilizing the Support Vector Machine (SVM) as a classifier results in superior performance compared to other classifiers.

Moreover, we propose a possible application of this model in conjunction with smart glasses technology to promote safe driving for individuals with diabetic retinopathy. By integrating the CNN model with smart glasses, drivers can receive real-time updates about their condition and any potential visual impairments that may affect their ability to drive safely.

This technology could prove to be a game-changer for diabetes patients, who often face challenges managing their condition while maintaining an active lifestyle. By combining advanced diagnostic tools with cutting-edge wearable technology, we can help ensure that individuals with diabetic retinopathy are better equipped to maintain independence, productivity, and quality of life.

References

1. Graveling, A.J., Frier, B.M.: Driving and diabetes: problems, licensing restrictions and recommendations for safe driving. Clin. Diab. Endocrinol. **1**, 1–8 (2015)
2. Kerr, D., Olateju, T.: Driving with diabetes in the future: in-vehicle medical monitoring. J. Diabetes Sci. Technol. **4**, 464–469 (2010)
3. Cox, D.J., Gonder-Frederick, L.A., Kovatchev, B.P., Clarke, W.L.: The metabolic demands of driving for drivers with type 1 diabetes mellitus. Diabetes Metab. Res. Rev. **18**, 381–385 (2002)
4. Sarki, R., Ahmed, K., Wang, H., Zhang, Y.: Automatic detection of diabetic eye disease through deep learning using fundus images: a survey. IEEE Access **8**, 151133–151149 (2020)
5. Ryu, K.S., Lee, S.W., Batbaatar, E., Lee, J.W., Choi, K.S., Cha, H.S.: A deep learning model for estimation of patients with undiagnosed diabetes. Appl. Sci. **10**, 421 (2020)
6. Chilukoti, S.V., Maida, A.S., Hei, X.: Diabetic retinopathy detection using transfer learning from pre-trained convolutional neural network models. TechRxiv, vol. 10 (2022)
7. Burewar, S., Gonde, A.B., Vipparthi, S.K.: Diabetic retinopathy detection by retinal segmentation with region merging using CNN. In: 2018 IEEE 13th International Conference on Industrial and Information Systems (ICIIS), pp. 136–142 (2018)
8. Tajudin, N.M., et al.: Deep learning in the grading of diabetic retinopathy: a review. IET Comput. Vision **16**, 667–682 (2022)
9. Doshi, D., Shenoy, A., Sidhpura, D., Gharpure, P.: Diabetic retinopathy detection using deep convolutional neural networks. In: 2016 International Conference on Computing, Analytics and Security Trends (CAST), pp. 261–266 (2016)
10. Yaqoob, M.K., Ali, S.F., Bilal, M., Hanif, M.S., Al-Saggaf, U.M.: ResNet based deep features and random forest classifier for diabetic retinopathy detection. Sensors **21**, 3883 (2021)
11. Vijayan, M.: A regression-based approach to diabetic retinopathy diagnosis using efficientnet. Diagnostics **13**, 774 (2023)

12. "kaggle" [Online]. kaggle.com/c/aptos2019-blindness-detection/data
13. Alwakid, G., Gouda, W., Humayun, M.: Deep learning-based prediction of diabetic retinopathy using CLAHE and ESRGAN for enhancement. Healthcare **11**, 863 (2023)
14. "Medium" [Online]. https://blog.devgenius.io/resnet50-6b42934db431
15. Alahmadi, M.D.: Texture attention network for diabetic retinopathy classification. IEEE Access **10**, 55522–55532 (2022)

Convolutional Neural Network and Long Short Term Memory on Inertial Measurement Unit Sensors for Gait Phase Detection

Mohammadali Ghiasi[1(✉)], Mohsen Bahrami[1], Ali Kamali Eigoli[1], and Mohammad Zareinejad[2]

[1] Mechanical Engineering Department, Amirkabir University of Technology, Tehran, Iran
Aradghiasi@Aut.ac.ir
[2] New Technologies Research Center (NTRC), Amirkabir University of Technology, Tehran, Iran
mzare@Aut.ac.ir
https://aut.ac.ir/content/197/Mechanical-Engineering

Abstract. Recent developments in robot exoskeletons have significant attention due to their ability to enhance mobility. Gait phase detection is a critical component of the control system for these devices, as it enables the identification of the current walking stage and facilitates appropriate ankle functionality. Gait phase detection has increasingly favored the use of inertial measurement units, which provide valuable data on angular velocity and acceleration. Traditional approaches, such as Hidden Markov Models, have been employed for gait phase detection. However, incorporating long-term dependencies into it can be challenging. An improved gait phase detection algorithm was presented in our article that exploits long-term dependencies by combining Convolutional Neural Networks with Long Short-Term Memory. The research aims to develop and evaluate models that can effectively incorporate previous values, enabling a better understanding of the temporal dynamics of gait. First of all, data from seven participants were collected using a gadget with an Arduino Nano 33 IoT attached to their feet. Performance evaluation was conducted using metrics, such as accuracy, precision, recall, and F1 score. The model successfully identified gait phases with an F1 score of 93.33%. The algorithm classifier's architecture enables fast adjustments to various gait kinematics using only one sensor attached to the foot. The findings of this study contribute to advancements in gait analysis, leading to improved control and adaptability of assistive devices, and enhancing rehabilitation therapies and prosthetic device development.

Keywords: Deep Learning technique · Gait phase detection · Inertial Measurement Unit

1   Introduction

Recently, there has been a significant surge in interest surrounding robotic exoskeleton technology, which has emerged as a rapidly advancing technology with extensive potential across diverse sectors, daily routines, and industrial environments. Among these applications, the lower limb exoskeleton has garnered considerable attention and is considered a valuable area of exploration in the medical field. This device holds immense potential in improving the mobility and physical function of patients during rehabilitation therapy, with the ultimate goal of enhancing their overall quality of life. Active prosthetic devices are designed to offer amputees increased comfort, safety, and fluidity of movements. Within the framework of the prosthesis control system, the primary device relies on the gait phase detection component for collecting data about the current walking stage. The technology for identifying walking patterns serves as a vital technical assurance for enabling the robot to analyze extensive sets of real-time data. Detecting the gait phase represents a commonly used method to visualize the position and stage of each patient [26]. However, for these devices to deliver a satisfactory level of performance, a functioning gait phase detection algorithm must be combined with them. After identifying the current phase, the prosthesis can provide the appropriate ankle functionality. Therefore, including an accurate gait detection approach specifically tailored to the prosthesis can significantly enhance the performance of the control system [2]. The Inertial Measurement Unit (IMU) provides valuable signal information regarding the angular velocity and acceleration of a person's gait. These devices have been effectively utilized in accurately predicting various events and phases of the gait. Consequently, when it comes to machine learning (ML) techniques, IMU signals are considered a favorable option. In the past, valuable approaches for event and phase detection have been proposed by several fellow researchers. There are two primary classifications for these computational methods. First, the threshold values are the simplest computing techniques for gait detection [21] and time-frequency analysis [11]. Second, ML algorithms are used for classifying offline and real-time gait stages. Manini et al. The utilization of Hidden Markov Models (HMMs) for estimating gait phases in sequence modeling has been extensively demonstrated. [14]. In sequence classification, each category is typically trained using a separate HMM, and Bayesian classification is utilized to determine the sequence's class [1]. However, incorporating long-term dependencies into HMMs to model the data can be challenging. Typically, this process entails redefining the state by incorporating previous values, resulting in expanded state space and increased complexity within the HMM. This is since HMMs only consider the current state when predicting the next state and do not have any recollection of previous states [10]. The achievement of Deep Neural Networks (DNN) in various domains such as image processing, natural language processing, and time series analysis has inspired researchers to utilize these networks in gait analysis [3]. Another unique method for estimating the percentage of the gait cycle is based on a DNN that incorporates IMU data which are installed on the lower leg. [22]. Several studies have presented impressive findings when conducted offline, although the effectiveness of real-time detection was rather unimpressive. This lackluster performance can be attributed to the time-consuming nature of the calculations, which is primarily caused by the large sizes of the parameter matrix [6]. The LSTM method is a viable option. Unlike other Recurrent Neural Networks (RNN) that experience the vanishing gradient problem when processing data with large temporal dependencies, the LSTM is unaffected by this issue [9]. To classify a broad range of activities through the use of wearable IMU sensors, Ordóez and Roggen (2016) employed a sophisticated CNN that was then followed by an LSTM [16]. By extracting features automatically from signals, the deep CNN was able to facilitate the detection of temporal correlations through the LSTM. This cutting-edge architecture allowed for the identification of gait phases online from time series. According to Ordóez and Roggen (2016), the issue of Human Activity Recognition (HAR) was tackled by employing a sliding window approach to classify activities, resulting in exceptional levels of accuracy. Nonetheless, the sliding window approach can be impeded by a delay in detection that may not be desirable for the control of corrective devices. This article aims to bridge the gap in gait phase detection algorithms by effectively capturing long-term dependencies. By addressing this challenge, the accuracy and reliability of phase identification can be enhanced, leading to improved control and adaptability of assistive devices. This, in turn, contributes to more effective rehabilitation therapies and advancements in prosthetic device development. To achieve this goal, the research focuses on developing and evaluating models that can effectively incorporate previous values, enabling a better understanding of the temporal dynamics of gait. By exploring innovative approaches and alternative models, the study aims to enhance the ability of gait phase detection algorithms to capture and utilize long-term dependencies. One promising innovation in addressing the challenge of long-term dependencies and improving gait phase detection performance is the combination of CNNs with LSTM. This architecture, known as a CNN-LSTM model, has demonstrated success in various domains and holds the potential for enhancing gait analysis. The integration of CNN and LSTM brings notable advantages in gait analysis. By leveraging the deep CNN's ability to extract meaningful spatiotemporal features and the LSTM's capacity to model long-term dependencies, this architecture significantly improves the accuracy and robustness of gait phase detection algorithms. Specifically, the research will explore innovative approaches, such as modifying existing models or proposing alternative models, to enhance the ability of gait phase detection algorithms to capture and utilize long-term dependencies. This could involve incorporating memory mechanisms or designing architectures that explicitly consider the temporal context of gait data. The evaluation of these models will involve assessing their performance in accurately detecting gait phases by considering various metrics. Additionally, the research will also investigate the impact of different input representations and feature engineering techniques on capturing long-term dependencies. The paper is organized as follows: First, a lightweight, inexpensive IMU sensor, which is implanted along the dorsal aspect of the foot without regard to orientation, is constructed and then

collected data. Secondly, the remarkable capacity of the CNN-LSTM classifier for recognizing transitions from kinematic data alone is achieved by continually updating the prediction during the ongoing phase. Finally, the influence that different hyperparameter settings have on the performance of the CNN-LSTM network is analyzed.

2   Experimental Setup and Data Collection

2.1   Experimental Setup

A gadget was constructed using an Arduino Nano 33 IoT with an integrated LSM6DS3 inertial sensor, which was placed inside a compact box designed using a 3D printer and equipped with a 9-volt battery. Data were collected from seven young participants using an IMU sensor. The embedded LSM6DS3 IMU sensor of the Arduino Nano 33 IoT was arbitrarily positioned and oriented on the participants' feet. The microcontroller (NANO 33 IoT) sampled the IMU data at 100 Hz (with a sampling time of 1 ms) and recorded the linear acceleration signals in the forward a_x and vertical a_z directions, as well as the angular velocity signal in the sagittal plane w_y, for classification purposes. This study established a wireless connection between sensor axes and Matlab/Simulink. The system was equipped with Intel Core i7-11700U CPU. We utilized the User Datagram Protocol (UDP) for rapid data transmission, prioritizing speed over data loss. This configuration facilitated real-time data analysis. UDP is preferred over Transmission Control Protocol (TCP) for this. It should be noted that UDP requires the Internet Protocol (IP) to be the underlying protocol for its operation [12]. The hardware is illustrated on the dorsal aspect of the right foot in Fig. 1.

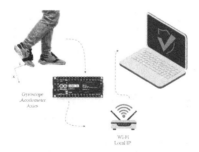

Fig. 1. Schematic of the experimental setup used to collect data on human walking motion with a single IMU on the foot dorsal.

2.2   Segmentation and Labeling

The gait cycle of walking can be divided into the stance phase and the swing phase. During the stance phase, certain events occur, including heel-off (HO), forceful plantar flexion, and toe-off (TO). In the swing phase, the foot can either experience a heel strike (HS) or be in a flat foot (FF) position. Understanding these phases is essential for comprehending the mechanics of walking. According to Perez et al. (2019), HS can be detected by observing the zero crossing in w_y after the positive peak during the swing phase on level ground. The timing of toe-off is not consistent and varies among individuals and strides. During level ground walking and stair climbing, toe-off occurs closer to the peak, while during stair descent, it happens closer to zero-crossing. An implementation of a complete gait phase model was carried out in the present document. The model was designed to take a data sequence as input, which is a small time series extracted from the IMU. To maintain the correct temporal connection between the data points in a gait phase, which was continuously recorded during the data collection phase. To remove inaccurate estimations, the steps were manually labeled. Figure 2 depicts labeled data collected while walking on level ground.

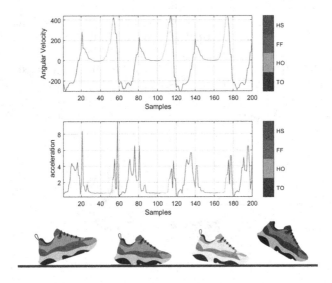

Fig. 2. The information that has been gathered is categorized according to the movements of the foot, including heel-off, toe-off, heel strike, and foot-flat. The process identifies four specific stages and converts the angular velocity of an IMU into an inertial frame.

3   Proposed Architecture

3.1   "Improved Sequence Learning with RNN and CNN"

RNNs have been devised to address the mentioned issue by integrating a recurrent link within each unit. Through this mechanism, neurons can effectively capture the temporal patterns present in sequential data. By receiving their activation with a weight and a time delay, neurons develop a memory of previous activations, enabling them to comprehend the dynamics of sequential input. This is achieved by feeding the neuron's activation back to itself with a time delay between interactions. Figure 3 visually depicts a single recurrent unit.

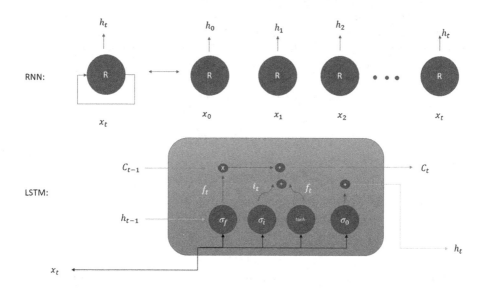

Fig. 3. Comparison between RNN and LSTM Unit

$$h_t^l = \sigma(w_{xh}^l z_t^l + h_{t-1}^l + w_{hh}^l + b_h^l) \qquad (1)$$

$$x_t^{(l+1)} = h_t^l w_{hy}^l + b_y^l \qquad (2)$$

In Eq. (1), the hidden state at time step t in layer l is denoted as h_t^l. The non-linear activation function is represented by σ, and b_h^l denotes the hidden bias vector. w_{xh}^l represents the input-hidden weight matrix, while w_{hh} stands for the hidden-hidden weight matrix.

Moving on to Eq. (2), w_{hy}^l indicates the weight matrix responsible for hidden-to-output connections, while the bias vectors are denoted by b terms.

The potential of these networks to achieve Turing completeness makes them theoretically suitable for sequence learning [20]. However, their memory architecture poses a challenge when it comes to learning real-world sequence processing [7]. To deal with this, LSTM was introduced as an extension of RNN. LSTMs utilize memory cells instead of recurrent units to store and retrieve information, facilitating the learning of temporal correlations across extended time spans. Gating mechanisms employed in LSTMs play a crucial role by individually defining the behavior of each memory cell. These mechanisms are based on component-wise multiplication of the input, enhancing the network's capability to capture and process sequential information. The LSTM updates its cell state when one of its gates is activated in response to that gate's input. The LSTM architecture incorporates multiple gates, such as input gates, output gates, and reset gates (also known as forget gates), to collectively determine the operations performed on the cell memory. Similar to RNNs, the activation of LSTM units follows Eqs. (3)–(7) [8]. The hidden state h_t of an LSTM cell is updated at each time step t. It is possible to represent the LSTM layer as a vector [16].

$$i_t = \sigma_i(w_{ai}a_t + w_{hi}h_{t-1} + w_{ci}c_{t-1} + b_i) \tag{3}$$

$$f_t = \sigma_f(w_{af}a_t + w_{hf}h_{t-1} + w_{cf}c_{t-1} + b_f) \tag{4}$$

$$c_t = f_t c_{t-1} + i_t \sigma_c(w_{ac}a_t + w_{hc}h_{t-1} + b_i) \tag{5}$$

$$o_t = \sigma_o(w_{ao}a_t + w_{ho}h_{t-1} + w_{co}c_{t-1} + b_o) \tag{6}$$

$$h_t = o_t \sigma_h(c_t) \tag{7}$$

LSTM is described by a set of equations. Equation (3) calculates the input gate activation (i_t) using the previous hidden state, previous cell state, and current input. Equation (4) computes the forget gate activation (f_t). Equation (5) updates the current cell state (c_t) by considering the previous cell state, forget gate activation, and modified current input. Equation (6) calculates the output gate activation (o_t). Finally, Eq. (7) computes the current hidden state (h_t) using the output gate activation and modified cell state. These equations allow LSTM cells to capture long-term dependencies in sequential data, making them valuable for various tasks like speech recognition and sentiment analysis. Recurrent neural networks can utilize raw sensor inputs as their input. However, incorporating attributes derived from the original sensor data often leads to improved performance [17]. Extracting meaningful features requires domain expertise, which limits systematic exploration of the feature space [5]. To address this, CNNs have been proposed as a solution [24]. A single-layer CNN captures features from the input signal through filter convolution or kernel operations. Each unit's activation in a CNN represents the outcome of convolving the kernel with the input signal. By utilizing convolutional operations to compute unit activations across multiple segments of the same input, it becomes possible to identify patterns stored by kernels independently of their position. During the supervised training phase of CNNs, the kernels are subjected to optimization to achieve the highest possible degree of activation for each subset of the class. An array (or layer) of units that all have the same parameterization is what is known as a feature map (bias and weight vector). The outcome of the kernel's convolution over the entire input dataset is determined by the activation. Convolution operator choice depends on data dimension: 2D kernels for image sequences [25], 1D kernels for temporal sequences like sensor signals [25]. In the 1D domain, a kernel filters data detects features, and eliminates outliers. Feature map extraction via 1D convolution is described in Eqs. (8) and (9) [16].

$$x_j^{(l+1)}(t) = \sigma(b_j^l + \sum_{f=1}^{F^l} k_{jf}^l * x_f^l(t)) \tag{8}$$

$$\sigma(b_j^l + \sum_{f=1}^{F^l} \sum_{p=1}^{p^l} [k_{jf}^l(p)x_f^l(t-p)]) \tag{9}$$

The activation $x_j^{(l+1)}$ of feature map j in layer $l+1$ can be calculated using Eq. 8. This calculation involves convolving the kernel filters k_{jf}^l with the feature maps x_f^l in layer l, followed by applying a sigmoid activation function. An alternative representation, as given by Eq. 9, incorporates a bias term and convolutions across multiple time offsets. Both equations encompass the essential computations necessary for transmitting information through the network layers. The dynamics and transformations occurring within the convolutional neural network architecture can be comprehended based on these equations, which serve as the foundation for understanding the process.

3.2   Efficient Hyperparameter Tuning for CNN-LSTM Classifier

The input signals are filtered and passed through a Rectified Linear Unit (ReLU) activation function with a threshold set at zero. To enhance the independence of translated features, a MaxPool layer is incorporated, extracting the maximum value within non-overlapping time frames of length L_P. The output of this CNN is a feature vector x_j, which is then input into the LSTM layer. This LSTM layer, consisting of N_L hidden units, effectively captures long-term temporal relationships. To generate probability distributions for each label, a dense layer with softmax activation is utilized on the feature vector x_j. The specific values of the structural parameters and their corresponding settings can be found in Table 1. The final layer of the LSTM-CNN architecture employs Softmax activation to classify the gait phase. By integrating these components, a probability vector is generated, indicating the likelihood of each class based on the extracted features from the previous. (See Fig. 4) This approach allows for the effective classification of input data by leveraging information from preceding layers. Furthermore, it ensures that the output is easily interpretable by other programs and applications [3]. The expression formula can be seen in Eq. 10. In this study, the employment of an LSTM-based model is crucial for the extraction of feature vectors from sensor data to make classification predictions at specific moments, which are determined by $t_p = L_P$. The frequency of categorization during each step is influenced by the LP parameter. The foot condition during the swing phase is promptly assessed by selecting the highest probability S_j With Softmax Equation (10) label at each time point.

$$S_j = \frac{e^{a_j}}{\sum_{k=1}^{N} e^{a_k}} \tag{10}$$

Here, a_j represents the parameter associated with event j, and N is the total number of events. The denominator ensures that the probabilities of all events sum up to 1, making it a valid probability distribution.

The performance of the CNN-LSTM classifier with hyperparameters $L_P = 1$, $L_C = 2$, $N_C = 32$, $N_L = 128$ was found to be effective across multiple subjects and gaits. Architect of the model in Fig. 4, which showcases the optimal performance of the classifier while simultaneously minimizing computational costs. These results emphasize the efficacy of the chosen hyperparameters and highlight the potential for practical implementation of the classifier in various applications. To evaluate the performance of the proposed CNN-LSTM model for gait phase detection, a one-subject-out cross-validation approach was employed. This approach involves training the model on data from all subjects except one and then testing it on the remaining subject. This process is repeated for each subject, ensuring that the model's performance is evaluated on data from different individuals.

The CNN-LSTM network was built using Matlab 2022b as the tool of choice. During the training process of the classifier, The Adam optimizer was used with a learning rate of 0.001, and categorical cross-entropy loss was employed as the loss function. A batch size of 2 samples was employed. After fifty epochs, there was only a negligible drop in the value of the loss function per epoch. Consequently, it was determined that the training process reached convergence after a total of 20 epochs.

Table 1. List of CNN-LSTM hyperparameters that can be adjusted

Hyperparameters	Define
L_P	Pooling length
L_C	Convolution kernels length
N_C	Filters are used in the convolution
N_L	Hidden units in the LSTM

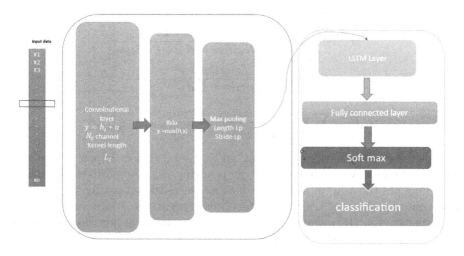

Fig. 4. The CNN-LSTM classifier extracts features, captures dependencies and generates class probabilities for time sequences [3].

4   Experimental Results

4.1   Performance Evaluation

A one-subject-out cross-validation approach was used to evaluate the model's performance for gait phase detection. The dataset was divided into seven subsets, and the model was trained on data from six participants and tested on the remaining participant in each iteration. This approach ensured robustness and effectiveness across different individuals. This evaluation approach helps validate the model's performance and its potential for real-world applications in gait analysis and assistive device control. In conclusion, the CNN-LSTM classifier with the specified hyperparameters demonstrates promise as an accurate and efficient tool for gait classification. Imbalances commonly arise when gathering gait phase data in real-world settings. Achieving highly accurate results is feasible by prioritizing the majority class and evaluating overall classification accuracy to gauge model performance. The assessment of classification system performance extends beyond relying solely on the total classification accuracy. The F-measure, commonly referred to as the F1 score, presents a more comprehensive evaluation by considering both false positives and false negatives. It integrates two fundamental metrics, precision, and recall in Eqs. (11), (12), which are defined based on the accurate recognition of total samples. Precision measures the proportion of correctly classified positive instances among all instances classified as positive, while recall captures the proportion of correctly classified positive instances out of all actual positive instances. Precision and recall, derived from the field of information retrieval, collectively contribute to the F-measure's robust evaluation of classification system effectiveness. As a result, the F1 score in Eq. (13) is typically a more meaningful performance metric than accuracy [23].

$$precision = \frac{TruePositive}{TruePositive + FalsePositive} \tag{11}$$

$$recall = \frac{TruePositive}{TruePositive + FalseNegative} \tag{12}$$

$$F_1 = \sum_i 2 * w_i \frac{precision_i \cdot recall_i}{precision_i + recall_i}, w_i = \frac{ni}{N} \tag{13}$$

Equation (11) presents a definition for precision. It defines precision as the ratio of true positives to the sum of true positives and false positives. Equation (12) describes the concept of recall, which measures the proportion of true positives out of the sum of true positives and false negatives. To assess the overall performance, Eq. (13) calculates the F1 score by combining precision and recall using weighted averages.

4.2 Results and Discussion

The CNN-LSTM model performs excellently in detecting gait phases, as shown in Confusion Matrices Fig. 5. For Precise classification results in Table 2 are achieved for the HS, FF, HO, and TO phases, with precision values of 92.43%, 90.86%, 96.05%, and 96.26% respectively. Moreover, recall rates of 85.35%, 95.02%, 92.90%, and 98.24% are attained for these phases. These outcomes underscore the CNN-LSTM model's ability to accurately identify and categorize gait phases. The F1 scores, which combine precision and recall, further validate its strong performance, reaching 88.75%, 92.89%, 94.45%, and 97.24% for HS, FF, HO, and TO respectively. The CNN-LSTM approach's effectiveness in gait phase detection can be attributed to several factors. Firstly, the CNN component facilitates robust feature extraction by learning distinctive spatial features from gait signals, enabling the model to capture essential patterns and characteristics associated with each gait phase. Secondly, the LSTM component leverages its sequential modeling capability to capture temporal dependencies inherent in gait signals, enhancing the model's classification accuracy by incorporating temporal dynamics. Through the synergistic combination of CNN and LSTM architectures, the CNN-LSTM model achieves a reliable and precise.

Table 3 presents a comparison of gait phase detection methods applied to lower limbs, allowing us to discuss and compare our method with other authors' works. Mannini [15] employed a single sensor and utilized an HMM to achieve an accuracy of 94% measured in terms of sensitivity for four gait phases. Evans and Arvind [4] utilized seven sensors and combined a Fully Connected Neural Network (FNN) with HMM, resulting in an accuracy of 88.7% (sensitivity) for five gait phases. Liu et al. [13] achieved an accuracy of 88.7 to 94.5% for eight gait phases using a NN approach with four sensors. Zhen et al. [26] employed a combination of LSTM and (DNN) with three sensors, achieving an F-score of 92% for two gait phases. Sarashar et al. [19] reported an accuracy of 99% with measured on the validation and training sets for three gait phases using an

LSTM model with a single sensor. Romijnders et al. [18] applied a CNN with two sensors and achieved a range of accuracy from 92% (Recall) and 97% (precision) for two gait phases.

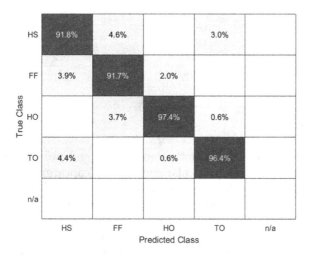

Fig. 5. Classification Confusion Matrix on the IMU, which was trained with six individuals and evaluated with one, approximately two minutes of typical walking time.

Table 2. Summary of classification performance of CNN-LSTM

	HS	FF	HO	TO	Average
Precision	92.43	90.86	96.05	96.26	93.9
Recall	85.35	95.02	92.90	98.24	92.87
F1	88.75	92.89	94.45	97.24	93.33

In comparison, our method utilizes a CNN-LSTM model with a single sensor, resulting in a notable F-score of 93.33% for four gait phases. This highlights the significant contribution of our approach in accurately detecting and distinguishing different gait phases. While some previous methods may exhibit higher accuracies for specific gait phases or rely on multiple sensors, our approach strikes a balance by achieving competitive performance with fewer sensors. These findings demonstrate the potential and practicality of our approach in real-world applications, offering an efficient and accessible solution for gait analysis and related fields.

Table 3. Gait phase detection methods applied on lower limbs.

Authors	Number of Sensors	Model	Phase Detection	Performance
Mannini	1	HMM	4	94% (Sensitivity)
Evans and Arvind	7	FNN-HMM	5	88.7% (Sensitivity)
Liu et al.	4	NN	8	Min 88.22 %, Max 94.5%
Zhen et al.	3	LSTM-DNN	2	92% (F-score)
Sarashar et al.	1	LSTM	3	99% (accuracy on train set)
Romijnders et al.	2	CNN	2	92% (Recall), 97% (Precision)
Our method	1	CNN-LSTM	4	93.33 (F-Score)

5   Conclusion

In conclusion, this study focused on the development and evaluation of gait phase detection algorithms, with a particular emphasis on capturing long-term dependencies. The experimental results demonstrated the effectiveness of the proposed CNN-LSTM architecture in improving gait analysis accuracy and robustness. The experimental setup involved an Arduino Nano 33 IoT with an integrated LSM6DS3 inertial sensor to collect data from seven participants. The data were segmented and labeled to identify different gait phases, such as heel-off, toe-off, heel strike, and foot-flat. According to the experimental results, the proposed CNN-LSTM classifier achieved a precision of 93.9%, a recall of 92.87%, and an F1 score of 93.33%. Overall, these results suggest that the CNN-LSTM classifier is a promising method for accurately classifying data. By leveraging the strengths of CNNs and LSTMs, the model demonstrated improved control and adaptability of assistive devices, leading to more effective rehabilitation therapies and advancements in prosthetic device development. Future research in this area could explore further modifications to the CNN-LSTM architecture, investigate the impact of different input representations and feature engineering techniques, and evaluate the model's performance on a larger and more diverse dataset. Additionally, real-time implementation of the CNN-LSTM model could be explored to assess its feasibility in practical applications. In conclusion, the integration of CNNs and LSTMs provides a promising solution for improving gait phase detection algorithms. This will improve assistance devices and rehabilitation therapies.

References

1. Bicego, M., Murino, V., Figueiredo, M.A.: Similarity-based classification of sequences using Hidden Markov models. Pattern Recogn. **37**(12), 2281–2291 (2004)
2. Cherelle, P., et al.: The ankle mimicking prosthetic foot 3-locking mechanisms, actuator design, control and experiments with an amputee. Robot. Auton. Syst. **91**, 327–336 (2017)

3. Coelho, R.M., Gouveia, J., Botto, M.A., Krebs, H.I., Martins, J.: Real-time walking gait terrain classification from foot-mounted inertial measurement unit using convolutional long short-term memory neural network. Expert Syst. Appl. **203**, 117306 (2022)
4. Evans, R.L., Arvind, D.: Detection of gait phases using orient specks for mobile clinical gait analysis. In: 2014 11th International Conference on Wearable and Implantable Body Sensor Networks, pp. 149–154. IEEE (2014)
5. Figo, D., Diniz, P.C., Ferreira, D.R., Cardoso, J.M.: Preprocessing techniques for context recognition from accelerometer data. Pers. Ubiquit. Comput. **14**, 645–662 (2010)
6. Flood, M.W., O'Callaghan, B.P., Lowery, M.M.: Gait event detection from accelerometry using the Teager-Kaiser energy operator. IEEE Trans. Biomed. Eng. **67**(3), 658–666 (2019)
7. Gers, F.A., Schraudolph, N.N., Schmidhuber, J.: Learning precise timing with LSTM recurrent networks. J. Mach. Learn. Res. **3**(Aug), 115–143 (2002)
8. Graves, A., Mohamed, A.R., Hinton, G.: Speech recognition with deep recurrent neural networks. In: 2013 IEEE International Conference on Acoustics, Speech and Signal Processing, pp. 6645–6649. IEEE (2013)
9. Hochreiter, S., Schmidhuber, J.: Long short-term memory. Neural Comput. **9**(8), 1735–1780 (1997)
10. Khalifa, Y., Mandic, D., Sejdić, E.: A review of Hidden Markov models and Recurrent Neural Networks for event detection and localization in biomedical signals. Inf. Fusion **69**, 52–72 (2021)
11. Khandelwal, S., Wickström, N.: Gait event detection in real-world environment for long-term applications: incorporating domain knowledge into time-frequency analysis. IEEE Trans. Neural Syst. Rehabil. Eng. **24**(12), 1363–1372 (2016)
12. Kumar, S., Rai, S.: Survey on transport layer protocols: TCP & UDP. Int. J. Comput. Appl. **46**(7), 20–25 (2012)
13. Liu, D.X., Wu, X., Du, W., Wang, C., Xu, T.: Gait phase recognition for lower-limb exoskeleton with only joint angular sensors. Sensors **16**(10), 1579 (2016)
14. Mannini, A., Genovese, V., Sabatini, A.M.: Online decoding of Hidden Markov models for gait event detection using foot-mounted gyroscopes. IEEE J. Biomed. Health Inform. **18**(4), 1122–1130 (2013)
15. Mannini, A., Sabatini, A.M.: Gait phase detection and discrimination between walking-jogging activities using Hidden Markov models applied to foot motion data from a gyroscope. Gait Posture **36**(4), 657–661 (2012)
16. Ordóñez, F.J., Roggen, D.: Deep convolutional and LSTM recurrent neural networks for multimodal wearable activity recognition. Sensors **16**(1), 115 (2016)
17. Palaz, D., Collobert, R., et al.: Analysis of CNN-based speech recognition system using raw speech as input. Technical report, Idiap (2015)
18. Romijnders, R., Warmerdam, E., Hansen, C., Schmidt, G., Maetzler, W.: A deep learning approach for gait event detection from a single Shank-Worn IMU: validation in healthy and neurological cohorts. Sensors **22**(10), 3859 (2022)
19. Sarshar, M., Polturi, S., Schega, L.: Gait phase estimation by using LSTM in IMU-based gait analysis-proof of concept. Sensors **21**(17), 5749 (2021)
20. Siegelmann, H.T., Sontag, E.D.: Turing computability with neural nets. Appl. Math. Lett. **4**(6), 77–80 (1991)
21. Vu, H.T.T., et al.: A review of gait phase detection algorithms for lower limb prostheses. Sensors **20**(14), 3972 (2020)

22. Vu, H.T.T., Gomez, F., Cherelle, P., Lefeber, D., Nowé, A., Vanderborght, B.: ED-FNN: a new deep learning algorithm to detect percentage of the gait cycle for powered prostheses. Sensors **18**(7), 2389 (2018)

23. Xia, K., Huang, J., Wang, H.: LSTM-CNN architecture for human activity recognition. IEEE Access **8**, 56855–56866 (2020)

24. Yang, J., Nguyen, M.N., San, P.P., Li, X., Krishnaswamy, S.: Deep convolutional neural networks on multichannel time series for human activity recognition. In: IJCAI, vol. 15, pp. 3995–4001, Buenos Aires, Argentina (2015)

25. Zeng, M., et al.: Convolutional neural networks for human activity recognition using mobile sensors. In: 6th International Conference on Mobile Computing, Applications and Services, pp. 197–205. IEEE (2014)

26. Zhen, T., Yan, L., Yuan, P.: Walking gait phase detection based on acceleration signals using LSTM-DNN algorithm. Algorithms **12**(12), 253 (2019)

Real-Time Mobile Mixed-Character License Plate Recognition via Deep Learning Convolutional Neural Network

G. Karimi[1], Z. Ali Mohammadi[1], S. A. Najafi[1], S. M. Mousavi[1], D. Motiallah[1], and Z. Azimifar[2(✉)]

[1] Department of Computer Science and Engineering, Shiraz University, Shiraz, Iran
[2] Deed Asia Development Group (17471), Shiraz, Iran
azimifar@cse.shirazu.ac.ir

Abstract. Automatic license plate recognition (ALPR) is one of the integrated parts of intelligent transportation system (ITS) technologies. In this work, an efficient ALPR algorithm has been developed to detect, differentiate, and recognize the Iranian national and free zone license plates (LPs), automatically and simultaneously. Latest versions of YOLO (you only look once) has been trained based on an in-house developed dataset for Iranian motor vehicles containing both national and free-zone LPs. In addition, an open-source multi-lingual OCR application has been trained to recognize alpha-numeric characters in the both LP types. Experimental results show that the generated ALPR pipeline can detect and recognize mixed characters in the LPs in real time and with high accuracy.

Keywords: Recognition · Detection · License plate recognition · Optical character recognition · OCR

1 Introduction

Rapid urbanization in recent years has led local authorities to recognize the increasing mobility needs of citizens as traffic volumes rise in urban areas on a daily basis. To eliminate the need for human intervention, automatic license plate recognition (ALPR) is frequently useful for studying the free flow of traffic and facilitate intelligent transportation [1]. In fact, ALPR has become a part of our daily lives and is expected to remain an integrated part of transportation technologies in the future. Nowadays, modern ALPR cameras not only read plates, but also provide other useful pieces of information such as the instantaneous number of vehicles, their direction, and speed. As a result, ALPR is finding its way into many aspects of the digital landscape. ALPR systems are being used without human intervention in a wide range of applications such as automatic access control, parking management, tolling, user billing, delivery tracking, traffic management, policing and security services, customer services and directions, red light and lane enforcement, queue length estimation, and many other services [2–8]. The ALPR process begins with the acquisition of high-quality license plate (LP) images from the intended

© The Author(s), under exclusive license to Springer Nature Switzerland AG 2023
M. Ghatee and S. M. Hashemi (Eds.): ICAISV 2023, CCIS 1883, pp. 77–93, 2023.
https://doi.org/10.1007/978-3-031-43763-2_6

scene using a digital IP camera. The captured images are then processed by a series of image processing-based recognition algorithms to obtain an alpha-numeric conversion of the LPs into a text entry [9]. LP configurations (styles, colors, fonts, sizes, etc.) vary widely, but the general process of an ALPR system is very similar and generally consists of three phases, LP detection, segmentation, and detection.

As seen in Fig. 1, LPs generally contain monolingual characters. However, in a few countries such as Saudi Arabia [10] and in some of the Iranian free zone regions such as Kish Island, LPs have bilingual characters as depicted in the figure. Specifically, Kish Island LPs have 2 × 7 numbers that are stamped in separate regions. There are also other signs and logos on the plates for regional identification. On the other hand, Iranian national plates contain a combination of 7 alpha-numeric characters, all of which are in Persian format. Considering Kish Island as a tourist attraction area, it is very common to see a mix of national and domestic vehicles with different LPs on the streets. This makes the ALPR process more complicated when compared to regular, national LP recognition.

The purpose of this research is to develop an efficient ALPR algorithm that can automatically detect, differentiate, and recognize Iranian national and domestic LPs. Conventional GPU-based object detection algorithms such as YOLO (you only look

(a)

(b)

Fig. 1. Diversities in (a) international LPs (b) national LPs

once) have been trained based on an in-house developed dataset for Iranian motor vehicles containing both national and free-zone LPs to detect the LP areas. The open-source, segmentation-free application PaddleOCR [11] has been trained using a dataset consisting of a mix of national and free-zone LPs to recognize alpha-numeric characters in the LPs detected by the YOLO-based detector in a single stage. The ultimate ALPR pipeline can detect and recognize mixed characters in the LPs in real-time and with high accuracy.

2 Related Works

An ALPR system is generally comprised of three stages: LP detection, character segmentation, and character recognition. The process starts with a digital camera capturing images of still or moving vehicles that contain LPs in the images. Depending on the image quality, the images may need to go through a series of image enhancing steps, such as denoising or tilt correction, to ensure that the images are suitable for character recognition in the next stages. Next, detect the vehicle using an object detection algorithm or its frontal view with the LP area. Then, the LP goes through a character segmentation step, which is based on the plate dimension and configuration, to identify alpha-numeric characters within the LP. This step has been omitted in some newer versions of ALPR algorithms, such as the one implemented in this study. This means that the character recognition is conducted in a single step. Many research efforts have been devoted to each of the main components of the ALPR systems and will be briefly reviewed in the following section.

2.1 License Plate Detection

Many authors have addressed the LP detection stage using variants of convolutional neural networks (CNNs). Montazzolli and Jung used a single CNN arranged in a cascaded manner to detect front views of car and LP regions and precision and recall used as the evaluation metric. They reported high recall and precision rates for their model [12]. Hsu et al. [13] adapted a CNN specifically for LP detection and showed that this new set of neural networks is efficient enough to run on edge platforms. Overall, the provided system has proven tolerant to extreme lighting changes such as angle changes and day/night, delivering competitive results when compared to state-of-the-art server hardware solutions. Achieved.

The YOLO (You Only Look Once) algorithm is an effective two-step technique that uses data augmentation like flipped characters and reversed number plates [14]. The CNNs which are used, are fine-tuned and trained at each stage of the algorithm and the trained model provides extremely fast and accurate results on wide variety of datasets. Sohaib Abdullah et al. used a complete LP recognition algorithm for vehicles in Bangladesh [15]. In this paper, YOLOv3 is used for detection of LP area. This area is uses for character localization and detection. The accuracy of digit and character recognition reported 92.7%. However, the application of their model is limited to the Dhaka metropolitan area. Harpreet and Balvinder have used fixed camera to study LP detection [16]. The captured video streams are converted into images. Edge detection and morphological operations are used to extract LPs area. For video stream with 240 and 200 frames the accuracy of 90.8% and 90% have reported, respectively.

2.2 Character Segmentation

License plates consist of a blend of alpha-numeric characters. Although the number of characters, color, font size and type as well as plate configuration are specific to each region, the space dimensions allocated to each character, the coordinates and the aspect ratio are normally fixed. This facilitates character recognition. Segmentation is performed by labeling the connected pixels in the binary LP image [17–22]. The labeled pixels with same aspect ratio and size of the characters are considered as LP characters. Hence, this technique is not applicable to extract all characters when characters are joined or broken. Bulan et al. [23] used hidden Markov models (HMMs). In this case, the most probable LP is determined by applying the Viterbi algorithm. They reported a very high accuracy LP recognition by jointly performing character segmentation and recognition.

Vertical edge detection was used with the long edge removal in [24, 25]. Nomura et al. used adaptive morphology approach to propose a segmentation algorithm for extracting severely degraded number plates [26]. Nomura et al. used the morphological thickness algorithm to find baselines and separate overlapping characters from each other. This approach detects base line and use it for segmenting connected characters. Segmentation rate of 84.5% was reported for 1000 image dataset from 1189 set of images [27].

2.3 Character Recognition

Template matching is a simple method for character recognition. In this method the characters which are extracted, are compared with a set of templates, Finally the template which has the most match with selected character is chosen. This method is usually used for binary dataset [9].

Comelli et al. [27] used template matching. This recognition process was based on calculating normalized cross-correlation values for all shifts of each character template on sub images containing LPs. It has been reported that more than 90% of the central processing unit (CPU) time was spent computing cross-correlation measurements between different templates and corresponding sub-images. Salimah et al. used OCR for character recognition of high-resolution images captured by Vivo 1610 smartphone. Application of the OCR method on the android operating system has resulted an overall accuracy of 75% [27].

3 Proposed Models

The real-time mixed-character LP recognition proposed in this work is performed in three stages; dataset preparation, LP detection, and character recognition. In the first step, vehicle datasets are prepared and the objects/classes are annotated. In the second step, the LP boundaries are extracted. The detection models YOLOv4 and YOLOv7 are used for detection purposes. The input size of the detection networks is set to 416×416

for YOLOv4 and 640 × 640 for YOLOv7. In the final stage, the recognition datasets are fed into Paddle-OCR model for training. Text detection and text recognition are two main parts in this approach, where the recognition part contains input images of different sizes. Minimum height and width of the recognition dataset are 36 and 166 pixels and the maximum height and width are 178 and 294, respectively. The pipeline for the present model is illustrated in Fig. 2.

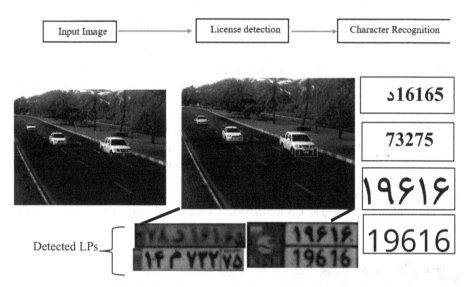

Fig. 2. Flow diagram for the present LPR algorithm

3.1 Data Preparation

A portion of the Microsoft COCO dataset is used to train the YOLOv7 model as our detector model. Since Microsoft COCO is a relatively large dataset that consists of eighty classes, most of which are neither related nor needed in our case of application, only a part of it in which at least one of the four license-plated classes (i.e., car, truck, motorcycle, and bus) existed, were selected using a python package named "Pycocotools". After the selection of the essential classes, LPs were added and labeled on the original portion of the COCO dataset. The selected portion of the COCO dataset that was used in the training process has 18,971 images in total. In order to have a model that would fit in our future application, we also utilized a dataset of 5,400 Kish Island images which were taken from traffic cameras located in different positions that gathered images from different times of the day, which not only enriched our training dataset with images similar to the ones that are going to be used on but also with images with different light exposure. The datasets were chosen carefully to have a model which has adequate generalization as well as good memorization. After training and testing the models, an expected pattern emerged; since traffic cameras have to track vehicles from a long distance to have an accurate speed measurement and a LP is a relatively small object to detect, our models suffered from low accuracy. In order to solve this problem, two more classes were added, "3LP" and "5LP". As the name tells us, they are labels with the same center as LP but have three and five times the width and height of the LP label because the recognition model has a separate detector model for itself, event detection of "3LP" or "5LP" would help to recognize LP while demonstrating a promising result. Figure 3 demonstrates two sample images from Kish Island and one from the Microsoft COCO dataset. For recognition, the model is trained on four different datasets from local cameras located in different places in Kish Island. The datasets consist of 7271 images in total with 1054 national and 6217 free-zone LPs.

As mentioned before, a mix of national and free-zone license plates is found in Kish Island. Therefore, the proposed automatic license plate recognition (ALPR) algorithms must be able to simultaneously handle both types of license plates in real-time. To do so, a dataset of combined license plates (national and free-zone) has been prepared from a local, fixed ALPR camera as well as other sources for the detector and recognition stages. For the detection stage, datasets have been labeled using the "ybat" labeling tool. Each of the vehicles in the dataset images has been carefully annotated (labeled) to specify the coordinates of bounding boxes surrounding the vehicles as well as their license plate areas. Annotation includes different vehicular classes with plates (e.g., cars, trucks, buses and motorcycles) and their license plates. Figure 4 illustrates the labeling process by the "ybat" tool. The labeled datasets are then used for training of the detection model.

Fig. 3. Typical examples of dataset images used for training detector

Once the detection dataset is prepared, the recognition datasets are labeled using an open-source tool named PaddleOCR. The national LPs are labeled using one bounding box representing Persian alpha-numeric characters and the free-zone LPs are labeled using separate bounding boxes representing both Persian and English numeric characters. An example of labeled LPs in this labeling tool is illustrated in Fig. 5.

3.2   Training Dataset for LP Detection

You only look once (YOLO) is used in this study as the object detector. The YOLO algorithm is a real time object detector by use of neural networks which has high accuracy and speed. A modern detector such as YOLO has two main parts, backbone and heed. Classes and bounding boxes are predicted by head part. Neck is additional layer in recent

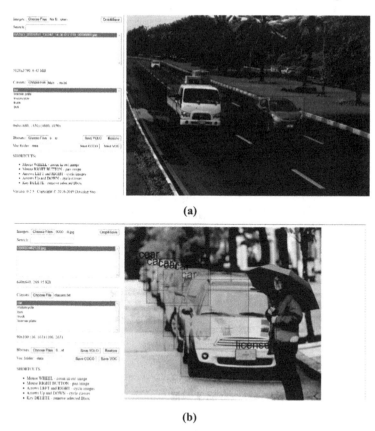

(a)

(b)

Fig. 4. Labeling different vehicular classes and their LP areas, (a) Kish Island sample data, (b) COCO sample data.

version of YOLO which is between backbone and head and is used to collect feature maps from stages.

In this study we used YOLOv4 [28] and YOLOv7 [29] to do the detection. These models also contain three stages. The YOLOv4 backbone is CSPDarknet53 [30] with 29 convolutional layers 3×3, a 725×725 receptive field and 27.6 M parameters. Over the backbone, an SSP [31] block with different maxpooling layer sizes of 5, 9, and 13 is added. Parameters from different backbone levels are aggregated by PANet method in YOLOv4. In the final stage, YOLOv3 [33] is used as the head. YOLOv7 [29], the newest version of YOLO has high speed and accuracy, 5 FPS to 160 FPS and 56.8%AP respectively. Yao Wang et al. made numerous changes in the YOLO architecture to achieve these results. The major changes in YOLOv7 include extended efficient layer aggregation networks (E-ELAN) [29].

Character Recognition Using PaddleOCR. PaddleOCR is an OCR framework or toolkit which provides multilingual OCR tools. PaddleOCR offers various models in its toolkit, among which the flagship PP-OCR is one of the best OCR tools available so far.

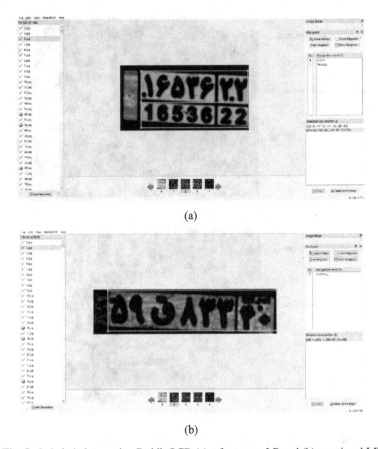

Fig. 5. Labeled plates using PaddleOCR (a) a free-zone LP and (b) a national LP.

As of now, PaddleOCR has three versions: PP-OCR, PP-OCRv2, and PP-OCRv3, all of which are ultra-lightweight [11, 34, 35]. In the present work, we have used the latest version, PP-OCR v3, since it is the most complete version of the toolkit for recognition stage [11].

Similar to other OCR technologies, PP-OCR v3 contains two main stages: text detection and text recognition which are for locating text in image and recognizing characters respectively.

Text Detection. The training framework in PP-OCRv3 detection model is collaborative mutual learning (CML) distillation [36], which is shown in Fig. 6. The main idea of CML is to combine the traditional distillation strategy of teacher guidance of students with Deep Mutual Learning (DML) [36], This allows the student networks to learn from each other. In PP-OCRv3, teacher and student models were optimized. PAN module with a large receptive field, LK-PAN (Large Kernel PAN) is used as teacher model [32] and the DML distillation strategy is adopted. The mechanism which is named RSE-FPN [11] is used for student model. Kernel size in LK-PAN is changed from 3×3 to 9×9 which causes easier detection of large fonts and extreme aspect ratios in texts. The structure

LK-PAN is shown in Fig. 7 and also DML for the teacher model, improves the accuracy of the text detection model. The structure of the student model in PP-OCRv3 is shown in Fig. 6, which uses a residual squeeze-and-excitation FPN (RSE-FPN). As we see in this figure, in RSE-FPN, the convolutional layer in the FPN is replaced by RSEConv [37] (Fig. 8).

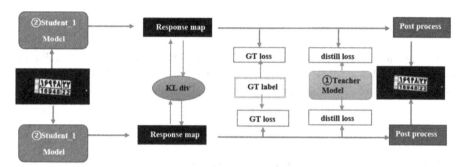

Fig. 6. CML distillation framework of PP-OCRv3 detection model [11]

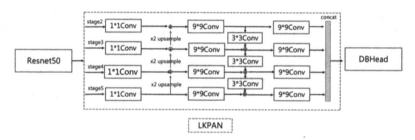

Fig. 7. The schematic diagram of LK-PAN.

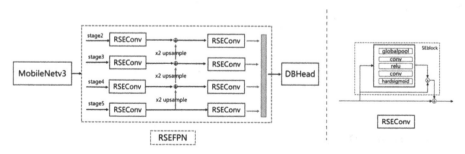

Fig. 8. The schematic diagram of RSE-FPN.

The total loss of detection part consists of 3 loss functions including the ground truth loss L_{gt}, peer loss from student model L_s and distill loss from teacher model L_t. The formula is shown in Eq. (1), where l_p, l_b and l_t are binary cross-entropy loss, Dice loss and L_1 loss respectively. The default values of α, β are 5 and 10 respectively.

$$Loss_{gt}(T_{out}, g_t) = l_p(S_{out}, g_t) + \alpha l_b(S_{out}, g_t) + \beta l_t(S_{out}, g_t) \qquad (1)$$

The peer loss uses KL divergence to compute distance between student models which is defined in Eq. (2):

$$Loss_{dml} = \frac{KL(S1_{pout}||S2_{pout}) + KL(S2_{pout}||S1_{pout})}{2} \qquad (2)$$

The final loss which refers to distillate loss. The distillation loss is shown in Eq. (3) where lp, lb are binary cross-entropy loss and Dice Loss respectively. And the default value of γ is set to 5. The f_{dila} is the dilation function which kernel is matrix [[1; 1]; [1; 1]].

$$Loss_{distill} = \gamma l_p(S_{out}, f_{dila}(T_{out})) + l_b(S_{out}, f_{dila}(T_{out})) \qquad (3)$$

Finally, the formula for loss function of the CML is as follows:

$$Loss_{total} = Loss_{gt} + Loss_{dml} + Loss_{distill})$$

Text Recognition. In recognition model of PP-OCRv3 the base model is replaced from CRNN [38] to SVTR (Single Visual model for Scene Text Recognition) [39]. The lightweight text recognition network SVTR-LCNet that its architecture is used in this version. This recognition model is proposed from SVRT-tiny by replacing the first half of SVRT-LCNet by PP-LCNet [40], reduce Global Mixing Blocks from 4 to 2 and moving it behind the pooling layer which is denoted as SVRT-G as shown in Fig. 9.

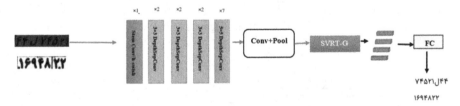

Fig. 9. SVTR LCNet architecture which is a light weight text recognition model

4   Results and Discussion

In this section experimental results and evaluation metrics are presented for the proposed models. Because models, datasets, and classes differ in recognition and detection, they are compared and presented in different sections. First, the dataset and then the models' results and metrics are demonstrated.

4.1 Detection

Because our test dataset is 50 images from Kish Island, which have high resolution and many objects to be detected, and the YOLOv4 model can't detect fine-grained objects, many of which exist in this case of study, the YOLOv7 base model is chosen as the detector model, due to the clear superiority of the YOLOv7 model compared to the YOLOv4 model in our experiments. YOLOv7 has six different sub-models and the base model which is used in our experiments is the sparsest. Despite being the sparsest sub-model and having less trainable weights compared to other sub-models, it shows promising results in both operation metrics and also frame per second (FPS) rate. Detector models are trained on a GTX 1080 GPU with 8 GB of GPU memory. Each epoch for training on the datasets takes almost 28 min.

Using the Microsoft COCO and Kish Island datasets for training and also 50 images consisted of Kish Island images which are taken in various places and having different light exposures, the result of the detection model with 5 and 7 classes are demonstrated in Table 1 and Fig. 10. Figures 11 and 12 demonstrate confusion matrices. In Table 1, the 7-class model lags in some of the performance measures compared to the 5-class model, but in the 7-class model, the recall for LP5 is 97.6% compared to 89.7% recall for LP in the 5-class model, which meets our expectations for the LP detection of far LPs, because LP3 and LP5 classes were added just to detect more LPs and having a more excellent recall demonstrates that finer-grained LPs are detected.

Table 1. Accuracy and run-time results of the proposed detector models on evaluation dataset

Models	Epoch	Class	Input image size	Prec.	Rec.	mAP 0.5
YOLOv7	221	5	640	0.89	0.96	0.97
YOLOv7	170	7	640	0.91	0.87	0.93

4.2 Recognition

The result of our training on recognition datasets is shown in Table 2. The name detection_2x refers to model which is trained with double-size LPs, we doubled the area of license plates (some parts of front view of car also selected) to get better result in the detection stage. As we see in this table, detection_2x has better results in detection than the first trained model.

Table 3 provides precision, recall and F1 for bilingual license plates. As we see in this table the PP-OCRv3 model performs well in recognizing of this Persian-English bilingual LPs. As we see in this table, a very high accuracy is achieved even with similar numbers (*i.e.,* ۱ and ۳).

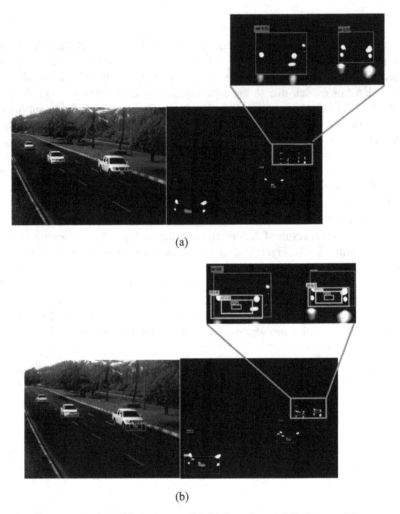

Fig. 10. (a) Detection with 5-class model (b) detection with 7-class model

To see the accuracy of our model on national datasets we have used evaluation datasets and prepared confusion matrix as shown in Fig. 13. For national datasets we see good results like bilingual datasets and also characters which are similar (*i.e.,* ب and پ) are recognized well.

In Fig. 14 we see some examples of recognition model for bilingual Persian-English and national images.

The OCR models are trained on a system with a Geforce RTX 3070 GPU, AMDRyzen 9 5900HX 33001 MHz CPU, and 32 GB of memory. The text detection and recognition in this stage are trained under 250 and 700 epochs, respectively.

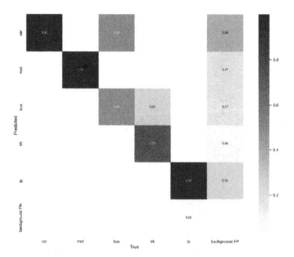

Fig. 11. Confusion matrix for the 5-class detector

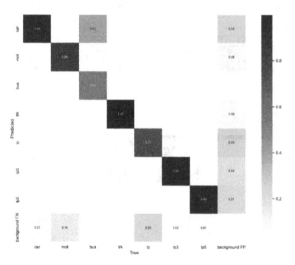

Fig. 12. Confusion matrix for the 7-class detector

Table 2. The results of recognition training

Stage	Precision	Recall	FPS
Detection	0.92	0.88	15.47
Detection_2x	0.95	0.92	8.56
recognition	0.97	0.97	955.68

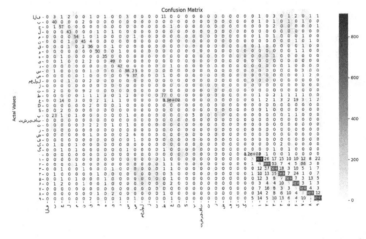

Fig. 13. The confusion matrix for national license plates with Persian characters and numbers

Table 3. The accuracy of each number in bilingual plates in Kish Island dataset

metrics	Numbers																	
	١	٢	٣	٤	٥	٦	٧	٨	٩	1	2	3	4	5	6	7	8	9
precision	0.99	0.99	0.99	0.98	0.98	0.97	0.99	0.99	0.99	0.98	0.99	0.98	0.95	0.97	0.95	0.99	0.93	0.97
Recall	0.98	0.99	0.98	0.98	0.99	0.98	0.97	0.98	0.98	1	1	0.97	0.97	0.91	0.97	0.99	0.97	0.96
F1	0.99	0.99	0.98	0.98	0.99	0.97	0.98	0.98	0.99	0.99	0.99	0.97	0.96	0.94	0.96	0.99	0.95	0.97

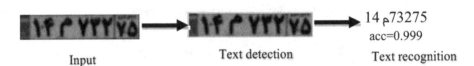

Input   Text detection   Text recognition

Fig. 14. Examples of recognition model

5 Conclusions

An efficient ALPR algorithm has been developed to detect, differentiate, and recognize
the Iranian national and free zone license plates (LPs), automatically and simultaneously.
A combination of YOLOv7 and multi-lingual PaddleOCR application have been trained

based on in-house developed datasets for Iranian motor vehicles containing both national and free-zone LPs. The resulting pipeline can efficiently and accurately detect and recognize the Iranian national and free-zone license plates. The detection and recognition accuracies are estimated at 95% and 97%, respectively.

References

1. Luo, X., et al.: Queue length estimation for signalized intersections using license plate recognition data. IEEE Intell. Transp. Syst. Mag. **11**(3), 209–220 (2019)
2. Lin, H.-Y., et al.: A vision-based driver assistance system with forward collision and overtaking detection. Sensors **20**(18), 5139 (2020)
3. Thangallapally, S.K., et al.: E-security system for vehicle number tracking at parking lot (Application for VNIT Gate Security). In: 2018 IEEE International Students' Conference on Electrical, Electronics and Computer Science (SCEECS). IEEE (2018)
4. Negassi, I.T., et al.: Smart car plate recognition system. In: 2018 1st International Conference on Advanced Research in Engineering Sciences (ARES). IEEE (2018)
5. Kanteti, D., Srikar, D., Ramesh, T.: Intelligent smart parking algorithm. In: 2017 International Conference on Smart Technologies for Smart Nation (SmartTechCon). IEEE (2017)
6. Shreyas, R., et al.: Dynamic traffic rule violation monitoring system using automatic number plate recognition with SMS feedback. In: 2017 2nd International Conference on Telecommunication and Networks (TEL-NET). IEEE (2017)
7. Chaithra, B., et al.: Monitoring traffic signal violations using ANPR and GSM. In: 2017 International Conference on Current Trends in Computer, Electrical, Electronics and Communication (CTCEEC). IEEE (2017)
8. Felix, A.Y., Jesudoss, A., Mayan, J.A.: Entry and exit monitoring using license plate recognition. In: 2017 IEEE International Conference on Smart Technologies and Management for Computing, Communication, Controls, Energy and Materials (ICSTM). IEEE (2017)
9. Mufti, N., Shah, S.A.A.: Automatic number plate recognition: a detailed survey of relevant algorithms. Sensors **21**(9), 3028 (2021)
10. Khan, I.R., et al.: Automatic license plate recognition in real-world traffic videos captured in unconstrained environment by a mobile camera. Electronics **11**(9), 1408 (2022)
11. Li, C., et al.: PP-OCRv3: more attempts for the improvement of ultra lightweight OCR system. arXiv preprint arXiv:2206.03001 (2022)
12. Montazzolli, S., Jung, C.: Real-time Brazilian license plate detection and recognition using deep convolutional neural networks. In: 2017 30th SIBGRAPI Conference on Graphics, Patterns and Images (SIBGRAPI). IEEE (2017)
13. Hsu, G.-S., et al.: Robust license plate detection in the wild. In: 2017 14th IEEE International Conference on Advanced Video and Signal Based Surveillance (AVSS). IEEE (2017)
14. Laroca, R., et al.: A robust real-time automatic license plate recognition based on the YOLO detector. In: 2018 International Joint Conference on Neural Networks (IJCNN). IEEE (2018)
15. Abdullah, S., Hasan, M.M., Islam, S.M.S.: YOLO-based three-stage network for Bangla license plate recognition in Dhaka metropolitan city. In: 2018 International Conference on Bangla Speech and Language Processing (ICBSLP). IEEE (2018)
16. Kaur, H., Aggarwal, M., Singh, B.: Vehicle license plate detection from video using edge detection and morphological operators. Int. J. Eng. Res. **1**, 1–5 (2012)

17. Kanayama, K., et al.: Development of vehicle-license number recognition system using real-time image processing and its application to travel-time measurement. In: (1991 Proceedings) 41st IEEE Vehicular Technology Conference. IEEE (1991)
18. Chang, S.-L., et al.: Automatic license plate recognition. IEEE Trans. Intell. Transp. Syst. 5(1), 42–53 (2004)
19. Nukano, T., Fukumi, M., Khalid, M.: Vehicle license plate character recognition by neural networks. In: Proceedings of 2004 International Symposium on Intelligent Signal Processing and Communication Systems, ISPACS 2004. IEEE (2004)
20. Shapiro, V., Gluhchev, G.: Multinational license plate recognition system: segmentation and classification. In: Proceedings of the 17th International Conference on Pattern Recognition, ICPR 2004. IEEE (2004)
21. Wu, B.-F., Lin, S.-P., Chiu, C.-C.: Extracting characters from real vehicle licence plates out-of-doors. IET Comput. Vision 1(1), 2–10 (2007)
22. Miyamoto, K.: Vehicle license-plate recognition. In: International Conference on Industrial Electronics, Control and Instrumentation (1991)
23. Bulan, O., et al.: Segmentation-and annotation-free license plate recognition with deep localization and failure identification. IEEE Trans. Intell. Transp. Syst. 18(9), 2351–2363 (2017)
24. Nain, N., Vipparthi, S.K. (eds.): ICIoTCT 2019. AISC, vol. 1122. Springer, Cham (2020). https://doi.org/10.1007/978-3-030-39875-0
25. Lin, N.H., Aung, Y.L., Khaing, W.K.: Automatic vehicle license plate recognition system for smart transportation. In: 2018 IEEE International Conference on Internet of Things and Intelligence System (IOTAIS). IEEE (2018)
26. Nomura, S., et al.: A new method for degraded color image binarization based on adaptive lightning on grayscale versions. IEICE Trans. Inf. Syst. 87(4), 1012–1020 (2004)
27. Comelli, P., et al.: Optical recognition of motor vehicle license plates. IEEE Trans. Veh. Technol. 44(4), 790–799 (1995)
28. Bochkovskiy, A., Wang, C.-Y., Liao, H.-Y.M.: Yolov4: optimal speed and accuracy of object detection. arXiv preprint arXiv:2004.10934 (2020)
29. Wang, C.-Y., Bochkovskiy, A., Liao, H.-Y.M.: YOLOv7: trainable bag-of-freebies sets new state-of-the-art for real-time object detectors. arXiv preprint arXiv:2207.02696 (2022)
30. Wang, C.-Y., et al.: CSPNet: a new backbone that can enhance learning capability of CNN. In: Proceedings of the IEEE/CVF Conference on Computer Vision and Pattern Recognition Workshops (2020)
31. He, K., et al.: Spatial pyramid pooling in deep convolutional networks for visual recognition. IEEE Trans. Pattern Anal. Mach. Intell. 37(9), 1904–1916 (2015)
32. Liu, S., et al.: Path aggregation network for instance segmentation. In: Proceedings of the IEEE Conference on Computer Vision and Pattern Recognition (2018)
33. Redmon, J., Farhadi, A.: Yolov3: an incremental improvement. arXiv preprint arXiv:1804.02767 (2018)
34. Du, Y., et al.: PP-OCR: a practical ultra lightweight OCR system. arXiv preprint arXiv:2009.09941 (2020)
35. Du, Y., et al.: PP-OCRv2: bag of tricks for ultra lightweight OCR system. arXiv preprint arXiv:2109.03144 (2021)
36. Zhang, Y., et al.: Deep mutual learning. In: Proceedings of the IEEE Conference on Computer Vision and Pattern Recognition (2018)
37. Hu, J., Shen, L., Sun, G.: Squeeze-and-excitation networks. In: Proceedings of the IEEE Conference on Computer Vision and Pattern Recognition (2018)

38. Shi, B., Bai, X., Yao, C.: An end-to-end trainable neural network for image-based sequence recognition and its application to scene text recognition. IEEE Trans. Pattern Anal. Mach. Intell. **39**(11), 2298–2304 (2016)
39. Du, Y., et al.: SVTR: scene text recognition with a single visual model. arXiv preprint arXiv: 2205.00159 (2022)
40. Cui, C., et al.: PP-LCNet: a lightweight CPU convolutional neural network. arXiv preprint arXiv:2109.15099 (2021)

Evaluation of Drivers' Hazard Perception in Simultaneous Longitudinal and Lateral Control of Vehicle Using a Driving Simulator

Mohammad Pashaee$^{(\boxtimes)}$ ⓘ and Ali Nahvi

Virtual Reality Laboratory, K. N. Toosi University of Technology, Pardis Street, Mollasadra Avenue, Vanak Square, Tehran, Iran
pashaee.gatabi@email.kntu.ac.ir

Abstract. Hazard perception is the driver's ability to detect and prepare for the proper reaction. Evaluation of hazard perception skills in the training and certification process is critical in reducing traffic accidents. Most hazard perception skill assessments are based on questionnaires and button clicks. In contrast, hazard perception can be more useful in practical driving based on realistic driving assessment criteria. In this paper, understanding driving motivations has been recognized as a key factor in predicting driver behavior. Physical variables such as time-to-collision, collision avoidance, and execution time have been employed to estimate numerical values for the motivations. A group of young drivers participated in driving simulator tests, and their behavior was assessed in terms of execution time, decision making, and decision execution. By investigating motivational parameters, the driver's behavioral anomalies are identified. Also, The drivers' hazard perception skills were thoroughly evaluated through simulated scenarios providing insights into their ability to perceive and respond to potential hazards in diverse traffic conditions. It was demonstrated that the algorithm could improve drivers' hazard perception skills in a risk-free simulator environment. If this algorithm is incorporated into driver's license training programs, drivers can improve their overall hazard perception abilities. This proactive approach ensures that drivers are equipped with the necessary skills and awareness to handle potential hazards on the road.

Keywords: Hazard Perception · Driving Motivations · Detecting Drivers Anomalies

1 Introduction

Driving is one of the most dangerous activities, involving many people directly or indirectly. Traffic accidents are also among the most crucial death reasons worldwide. In quest of saving lives, it is of high significance to propose prevention solutions to the high casualty statistics [1]. Therefore, training drivers to diagnose and to seriously take the proposed risks could be the primary strategy to reduce traffic accidents.

The first and most significant study regarding high-risk driving behaviors was conducted by Reason et al. [2]. According to the mentioned study, high-risk driving behaviors consist of slips, errors, and, violations. Slips are deviations, commonly occurring as a result of attention, memory, and information misprocessing. Errors occur when the intended actions fail to produce the expected outcomes, and Violations are behaviors contravening the essentials of safe driving.

Studies in developed countries have shown that identifying the human behaviorial factors and using educational approaches to eliminate and rectify driving behavior, plays an essential role in preventing traffic accidents. Also hazard perception deficiency has a significant impact on accidents. In this regard, many governments take traffic hazard perception tests as part of the driving license procedure for applicants; In the UK, traffic hazard perception has also become a part of theoretical driving tests since 2002 [3].

Sagberg and Bjornskau [4] provided a practical definition of hazard perception skills and showed that potentially dangerous response times consist of two stages: perception and reaction. It is also assumed that any change in the observed reaction time is essentially related to the perceptual phase. Thus, hazard perception has two distinct components; known to be the amount of perceived danger in each situation and the perceived reaction time in danger. In other words, not only the identification and prediction phases of the danger is crucial, the ultimate response to the imposed danger also plays an important role. In this regard, Grayson [5] proposed a four-component model of hazard response: A) hazard identification and determination accuracy: knowledge and awareness that the danger may arise. B) threat assessment: evaluate the situation and assess whether the hazard is significant enough to require a response. C) response selection: choosing an appropriate response among available decisions. D) Action: implementing the selected response.

Machin and Sankey [6] have studied the behavioral characteristics of young Australian drivers. Results indicated that the inexperienced drivers are incapable of estimating the proposed risk. Borowsky [7] published an article on hazard perception, analyzing and comparing the reaction of inexperienced young drivers with that of the experienced drivers toward pedestrians on inner and suburban routes. In the test, the eye movements of the drivers were tracked to obtain their movement structure; the data were then applied to calculate the drivers' focus time on the pedestrians.

Another study [8] compared two different approaches: The motive consisted of thirteen animated video clips containing hidden dangers. These hidden dangers were either visible driver/pedestrian who suddenly starts a risky activity due to a particular circumstance or visible driver/pedestrian in the course of collision. In the first part, the participants are supposed to identify the most important hidden dangers after watching the clip; while in the second part, they are asked to identify the danger immediately. In both sections, scores were given based on the number of significant hidden hazard identification. Skilled participants scored remarkably better in both sections in comparison to young drivers.

The relationship between each driver's characteristics and behavior was investigated using a demographic questionnaire and the Manchester driving test. Then, variance and correlation methods were applied to analyze the obtained data. Finally, research showed that the stress index, the level of errors, and logical actions are all directly and significantly related [9].

Studies have shown that the use of images, videos [10], and simulators [11] are among the primary methods for enhancing driver hazard perception skills. When comparing these methods, the use of simulators is considered more appropriate because the driver will be placed in similar conditions to real-world situations. Additionally, in most experiments, the researcher's hazard perception criteria are either the questionnaire [12] filled by the driver or the animated video test; however, neither of the mentioned criteria can be considered a precise assessment method. Also, some researchers use the reaction time of the driver [13], and physical or biological measures [14], however, even this parameters is not able to characterize the perceived danger and propose a solution to eliminate the anomaly.

If the main factors regarding the behavior selection are numerically estimated, the driver behavior and hazard perception could be adequately analyzed. Therefore, it is possible to diagnose the driver behavior abnormalities and evaluate whether the hazard perception skills are used accurately. The aim of the present study is to **determine and estimate the effective factors of the driving hazard perception, in order to examine behavioral anomalies of the driver with the aid of a driving simulator in which longitudinal and lateral performance of the vehicle are taken into consideration.**

2 Hazard Perception

Driving hazard perception is the ability to detect a dangerous or possibly accidental leading situation [15]. Potentially hazardous situations are the ones in which the driver is obligated to make a quick change in speed or direction to avoid a collision. Identifying and predicting dangerous situations while driving is an important skill, allowing the driver to overcome complex cognitive demands in traffic space [16]. Hazard perception consists of hazard tracking and hazard assessment, followed by selecting and implementing an appropriate response.

It is best to know the driving cycle to determine the hazard perception. Many researchers concluded that the driving process could be composed of behavioral components such as observation, environmental cognition, hazard perception, emotional adaptation, decision making, and decision implementation [17]. The driving cycle is shown in Fig. 1.

According to Fig. 1, driving is affected by hazard perception and emotional interactions, consequently influencing all major driving components. Therefore, it is necessary to determine which component or components are experiencing any form of anomaly; therefore, hazard perception anomalies are classified accordingly. The following examines the driving components to introduce an appropriate classification of hazard perception disorders.

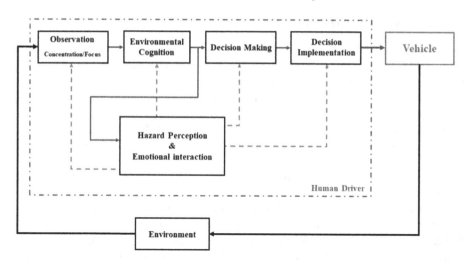

Fig. 1. Driving Cycle.

2.1 Observation

In the present study, observation implies collecting the physical information from the driving environment at a specific time and space. In other words, the observation is repeated frequently, using variable frequencies. The environmental sampling frequency in the observation process generally depends on the driver's concentration. Similarly, the detail received by the driver is also related to his/her focus on that particular space. Therefore, the characteristics of concentration and focus express the observation quality of the environment.

Hazard perception motivates the driver to identify and observe the risk factors with more awareness and concentration. This means less processing is assigned to other factors, therefore, the driver might not be able to observe other proposed risks. So, hazard perception can cause anomalies in the observation process. On the other hand, the driver may allocate his/her computing space to other mental tasks and may not observe dangerous factors correctly, which means the hazard perception is hindered due to observational errors. Accordingly, the following anomalies are defined to study the hazard perception and observation:

- Excessive focus and attention due to hazard perception.
- Lack of hazard perception due to observational errors.

The first anomaly causes error in decision-making; the driver considers fewer risk factors, as the criteria for decision making, while riskier factors should be regarded dangerous. The second anomaly generally leads to an initial time delay and will increase the reaction time.

2.2 Environmental Cognition

Decision-making with the aid of physical information, obtained by the observation process, could be very complex. Combining the physical data, which include lots of information, enables the driver to estimate meaningful physical variables and therefore provide suitable measures for decision making. These concepts and variables are considered as determinants of the behavioral factors. According to the mentioned issues, the environmental cognition component calculates the determinative behavioral factors based on the recorded physical data, providing the required components of hazard perception and decision-making.

Obviously, the hazard perception could be obtained by comparing the determinant behavioral factors with the values expected by the driver. When the driver feels threatened, his desired amount is reduced, feeling more threatened. If the driver miscalculates or improperly chooses the determinative behavioral factors, inappropriate hazard perception will happen, resulting in inadequate emotional interaction. In some cases, the feeling of danger in the last moments could cause the driver to exaggerate or underestimate the determinative behavioral factors, which will lead to errors in both decision making and even decision implementation. The following two anomalies could be defined in terms of hazard perception and recognition:

- The inadequate determinative behavioral factors in the cognition component could cause hazard perception errors.
- Hazard perception could cause the reduction or exaggeration in estimating the determinative behavioral factor in the cognition component.

2.3 Decision Making

The driver decides based on the determinative behavioral factor values in the decision-making component. Generally, this question arises: on what basis does the driver make the decision? In traffic psychology, three different theories have been proposed regarding the human decision-making mechanisms.

In the first theory, known as the control theory, the intelligent agent receives feedback from the momentary value, compares it with the desired value, and calculates the error. Then, the intelligent agent tries to lower the error down to zero and set the quantity value equal to the desired value. According to psychologists, one seeks self-satisfaction in driving, and therefore, no attempt is in reducing the functional error to zero is commonly observed. The mentioned point of view is not recommended by the humanities experts, however, control engineering experts inevitably use this theory in numerical modeling to present driver behaviors in terms of transfer functions or controllers [18].

Some traffic psychologists [19, 20] have proposed a behavior balancing mechanism factor within a specific range instead of a particular value. This balance, also seen in body physiology, is known as the Hemostasis theory which states that many decisions placed within a permitted area have the same choice priority and mathematically speaking, do not lead to a specific answer; this issue could cause significant problems in numerically modeling the driver's behavior.

Some believe that when the determinative behavioral factor is in the allowable range, fewer priority factors could be used to optimize the answer in the permissible range and reach a specific unique decision point. In other words, from the optimization point of view, the Homeostasis balance theory is regarded as optimization constraints, and second-order factors are considered as cost functions; therefore, the driver is a constrained optimizer of the determining factors [21, 22].

Hazard perception affects the permitted range of determination and driving safe area. If the hazard perception exaggerates in limiting the restricted areas, the driver would make a mistake in choosing the right decision or will not make any decision; since the permitted area is zero. On the other hand, bad decision-making causes the driver to avoid hazard factors and feel greater danger. Generally, hazard perception and decision making anomalies are categorized as follows:

- Hazard perception imposes more limitations on decision constraints resulting in confusion.
- Bad decision-making intensifies the hazard perception in the future.

2.4 Decision Implementation

In this component, the driver implements the decisions made during the decision-making step. Similar to proportional controllers in control engineering, the driver adjusts the actual values by considering the proportional control gains. Hazard perception could affect the control values. The driver needs more time to control if the proportional gains are low. If the values are high, the control process will experience an overshoot, or it may become unstable.

Describing the driving components indicated that hazard perception and emotional interaction affect all major components. On the other hand, hazard perception is highly dependent on determinative behavioral factors. The proposed hazard perception models are not able to evaluate these effects correctly to solve the anomaly issue. If determinative behavioral factors are estimated based on physical road information in any traffic situation, it would also be possible to assess the driver's expected danger value. Therefore, the performance of the driver regarding the hazard perception, based on determinative behavioral factors, would also be determined. Then, using the performance analysis, it would be possible to determine the anomalies between the hazard perception and other major driving components. With this method, there would be no need to assess hazard perception of the driver based on questionnaires, button-pushing, or other fundamental techniques.

3 Driving Motivations: Determinative Behavioral Factors

The driver provides proper decision-making criteria, using meaningful concepts and variables containing lots of information. These criteria are named driving motivations. These motivations measure the efficiency of each possible decision in each traffic situation. Numerous psychological theories have attempted to define the driver's determinative behavioral factor.

Based on Task difficulty hemostasis theory, the driver tries to keep his behavior within a specific range so that the task difficulty does not exceed the allowable limit [20]. Task difficulty is based on two concepts of demand and capability. Capability indicates the response performance that the driver is able to provide for the task. This quantity is related to the main driver characteristics, including experience, training, practice, competence, and human factors. Unfortunately, these factors cannot be measured in real-time; however, there is a common dominator. All biological abilities and insight skills affect the reaction time of the driver. In fact, the reaction time against traffic accidents is the resultant of the biological and behavioral abilities. Task difficulty is estimated based on environmental conditions, other road user behaviors, vehicle conditions, speed, human factors, location, and finally, the road trajectory. The impact of all these parameters on the task demand lead to complexity. Similar to the driver capability concept, the required reaction speed could be affected by the environmental condition, road trajectory, and other behaviors of the deriver.

Gibson and Krug introduced the safety margin theory in 1938, which is based on the concept of the safe navigation field [23]. The theory defines driving as a time-space process and describes a safe space as a dynamic area in front of the car. Safe zone boundary is determined by fixed and movable obstacles, passage edges, and dynamic and momentary vision fields of the driver. Considering the extra time for the operation, the driver tries to create a safe margin based on traffic condition complexity and the possibility of unwanted obstacle movements [24]. From a control engineering point of view, the safety margin is a measure of uncertainties. The driver is meant to maintain a secure distance from unsafe boundaries to take the perceptual errors, unwanted movements, and vehicle dynamic behavior into account correctly, therefore, the safety margin criteria determine the motivation of the driver to face uncertainties. The safety margin could also be considered a decision constraint or a factor that the driver tries to maximize.

The cognitive control model considers the time factor effect on the driving cycle. According to the model, the driver behaves in a way that the time required for perception, decision-making, and action is less than the total usable time; this relation is considered an unequal constraint on the behavior of the driver [25]. The theory is the basis for defining the concept of affordability. The consumable time is usually the collision time in driving. If the execution time is less than the collision time, the decision could be considered as applicable. On the other hand, if the execution time, which is the summation of perception, decision making, and implementation, is longer than the collision time, the decision is unaffordable. According to the mentioned discussions, determinative behavioral factors are:

- Collision avoidance difficulty motivation is calculated by comparing task demand and the capability of the driver.
- Motivation to prevent the tire from slipping.
- Motivation to compensate for the impact of uncertainties (safety margin).
- Motivation to measure the unaffordability.

Task difficulty can be determined using the cognitive control time cycle. If the decision puts the vehicle in a situation where the collision time is less than the time required for collision avoidance, the unequal time condition is violated, on the other hand, these conditions are indicators of task difficulty. Therefore, based on the cognitive control time cycle approach, the concept of task demand could be related to Generalized Time to Collision (GTTC), which is developed in [26], that is, the traffic conditions inflict their effect on the collision time. The shorter the collision time, the faster performance for traffic conditions and higher demand for duty is required, therefore, task difficulty criteria could be estimated by comparing the collision speed, as the task demand, with the driver reaction time as the driver ability.

Similarly, the safety margin is related to the time difference between time to collision and driver collision avoidance time. In fact, the safety margin is related to the definition of safe driving and safety boundaries, therefore, it is generally defined based on duty demand and driver ability. Since these motivations are estimated using both velocity and direction, safety margin could also be defined for each. It should be noted that defining the safety margin in a two-dimensional space of speed and direction can also define safety boundaries in terms of velocity and direction. Therefore, the concept of safe boundary distance could be defined in terms of directional velocity distance. Accordingly, the safe margin means that the driver is away from the speed limit defined by the safe border and can change the vehicle speed in the event of accidents or other uncertainties. Similarly, the driver must be away from the safe zone direction and be able to change the direction in the event of uncertainties.

The motivation criteria to keep the car stable results from the fear of slipping. The transverse force in tires is a function of central directional acceleration. The lateral acceleration to the center is also a function of radius curvature and vehicle velocity; lateral tire force is related to the direction and velocity and could be calculated for each pair of velocity and direction. Therefore, the directional acceleration affects the risk perception after the driver chooses a decision and implements the specific direction and speed.

4   Detecting Algorithm of Driving Behavior Anomalies

The present project aims to assess the driver hazard perception to identify the anomalies regarding the driving behavior. The detecting algorithm is developed to serve this purpose. Figure 2 schematically illustrates the detecting algorithm of driver behavioral anomalies; this algorithm is able to identify the driver behavioral anomalies causing the collision. Driver hazard perception analysis requires comparing the driver behavior with standard criteria. Therefore, the hazard perception function described below has been developed to investigate the driver behavior and assess the algorithm applicability. The function is used in two modes: the first mode implies the function with the driver delay time, and in the second mode, the function acts as an intelligent driver assistance system in the moment of danger, taking control of the vehicle.

According to Fig. 2, the algorithm first evaluates the driver observation component. If the driver has a perceptual error due to the visual misunderstanding, the anomaly is observed as reaction time increases; this means that the driver is too late. Also, suppose the hazard perception function cannot avoid the collision with the driver delay. In that case, the driver delay could be considered the cause of the collision, stating that the driver had a perceptual anomaly over the reaction time. If the hazard perception function is able to prevent a collision, even with driver delay, then the collision cannot be attributed to the reaction time.

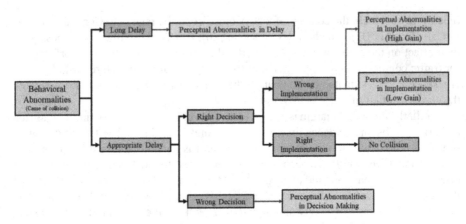

Fig. 2. Detecting algorithm of driver behavioral anomalies.

If it is specified that driver delay is not the collision cause, the decision will be compared to the second hazard perception function. If the driver chooses a different decision and the collision is taken place, the collision stems from the decision made by the driver, that is, the driver has not chosen the right decision among available options by miscalculating the driving motivations. Therefore, the perceptual anomaly in the decision-making is observed in his behavior. However, if the decision is equal to the second hazard perception function decision and a collision occurs, a more careful analysis of the driver's behavior should occur.

In a dangerous traffic situation, incorrect motivational estimations not only lead to wrong decisions but also would influence the decision implementation. So if the driver has made a decision similar to the hazard perception function but is still unable to avoid the collision, the cause could be found in the decision implementation. In fact, the driver has a behavioral disorder in the decision implementation component. As mentioned earlier, the driver acts as a proportional controller, so choosing the proper control gain is important. If miscalculations cause the driver to choose the wrong gain, trouble would arise in decision-implementation. Anomalies in the decision-implementation component could occur in two forms, in the first case, the driver miscalculates the severity of the situation and engages in less severe behavior than required. The drivers' gain to avoid collision has been small. In the other case, the driver exaggerates the situation by magnifying difficulty in his motivations, eventually leading to choosing a more effective gain.

4.1 Hazard Functions

The Hazard perception functions are based on a driving assistant model [26] has a different attitude and structure from previous robotic and control engineering research in the field of path design. Unlike previous models [23, 27], the presented model does not design any routs to avoid collision and reach the destination but chooses the best decision among all paths and curves based on driving motivations to meet the driving criteria. A measurable variable of physical information among traffic arrangements is selected so that a numerical estimate could be obtained. In this model [26], the ability of the driver was considered as one of the motivational factors; The situation complexity would require the driver to spend time processing the information. If the traffic conditions become complicated, the driver will need more time to process, therefore, reducing the speed prolongs the collision time and increases the usable time for decision-making. As mentioned earlier, the collision time indicates how fast the driver has dealt with the obstacle. The concept is comprehensible to humans and, in the meantime, creates a feeling. It also expresses the severity, in case of a collision with the obstacle. Thus, the conditions in estimating the required demand are inversely related to the time to collision. The shorter the time to collision, the more ability the driver needs for decision-making and data processing. In fact, the task demand model is assumed to be proportional to inversed collision time, and the ability of the driver is proportional to inversed usable time or collision avoidance time. Hence, the estimation of task demand (D) and driver capability (C) is carried out in the following manner [26]:

$$D = \frac{1}{TTC} \tag{1}$$

$$C = \frac{1}{TTA} \tag{2}$$

where the time to collision (TTC) is estimated using a first-order approximation, and the time to avoidance (TTA) is calculated as the sum of the driver's time delay and the vehicle's braking time.

The model states that a decision is unaffordable when the path curvature and speed values are different from current speed and path curvature values of the vehicle, and the available time for decision implementation is short, that is, the models use the comparison between the action and the usable time as the unaffordability criteria as follows [26]:

$$U = \left(\frac{1}{AT} - \frac{1}{UT} \right) \tag{3}$$

where the estimation of usable time (UT) is based on the time to collision of the current speed-steering angle, while the action time is determined by considering factors such as vehicle acceleration time, braking time, steering time, and driver time delay.

To avoid slipping, the model analyzes the tire slip force in exchange for force variations which means that the driver tends to keep the vehicle in a stable area.

Finally, the model considers the safety margin for the two parameters of velocity and curvature. In other words, it determines how far is the current curvature and speed from the safe boundary speed and curvature. Using these estimations and considering the decision-making action as a minimization problem, the model chooses the best decision. In order to imply the decision, the model adjusts the current curvature radius and velocity to the one decided by the model.

Hazard Perception Delay Function

In the first case, the model is considered with a driver delay; that is, until the driver's first reaction, the hazard perception function is not activated; however, after the driver's delay time, the model will choose the best decision and avoid the collision. Therefore, the hazard perception function condition considering the driver delay is as follows [26]:

$$if \begin{cases} Time < \tau_d \rightarrow Active_d = 0 \\ Time > \tau_d \rightarrow Active_d = 1 \end{cases} \tag{4}$$

In the above relation, Time is the test time, τd is the driver reaction time, and Actived is the hazard perception function activated with the driver's delay. The purpose of designing the hazard perception delay function is to investigate how the driver behaves to achieve the perceptual anomaly in time. According to the detecting behavioral anomalies algorithm, if the hazard perception function of the delay could not successfully avoid the collision, the delay is responsible for the collision.

Driver-Assist Hazard Perception Function

The second hazard perception function examines the current situation along with the performance of the driver; the function is activated whenever the driver cannot avoid the collision, taking control of the vehicle. The hazard perception function requires the following conditions in order to be activated:

1. If the task demand for the current curvature radius and speed of the vehicle is greater than the maximum driver capability (C_{max}), the hazard perception function activates the system.
2. In case the estimated tire force, due to the curvature radius and velocity, is above the transition zone and the vehicle becomes unstable, the hazard perception function activates the steering system.
3. The hazard perception function must activate the system before curvature and speed unaffordability exceeds zero.
4. Before the safety margin value of the decision curvature and speed reduces to lower than the minimum value (S_{min}), the hazard perception function must activate the automatic steering system.

According to the proposed conditions, the activation of the collision avoidance system by the hazard perception function is expressed as follows [26]:

$$if \begin{cases} D(\kappa^k, v^k) > C_{max} \ or \ \left(f_{y_r} \geq f_{r_2} \cap f_{y_f} \geq f_{f_2}\right) \\ or \ U\left(\kappa_{dec}^{k+1}, v_{dec}^{k+1}\right) > 0 \ or \ S\left(\kappa_{dec}^{k+1}, v_{dec}^{k+1}\right) < S_{min} \end{cases} \tag{5}$$

In the above relation, D indicates the task demand, U is unaffordability, S is the safety margin, f is the tire lateral force, and active indicates the command to activate the collision avoidance system by the hazard perception function. Regulating the value of these conditions affects the hazard perception function sensitivity to interfering with the driving behavior. Since the hazard perception function aims to investigate the driver anomalies regarding the decision-making and implementation, its behavior, as a criterion, should be appropriate. Therefore, the hazard perception activation values are chosen so that the best decision is observed after the activation in different traffic configurations. Therefore, $S_{min} = 0.1$ is considered the activation condition from the safety margin point of view, while $C_{max} = 1(1/s)$ is considered as an activating condition for the hazard perception function as the task demand point of view. The maximum value for the task demand is that the hazard perception driver assistant function must be activated whenever the collision time is less than 1 s.

5 Scenarios

As shown in Fig. 3, the driving simulator investigates the driver hazard perception anomalies. In the design of this simulator, the front half of a Pride vehicle, along with all its components such as the dashboard, console, and other interior parts have been considered. From the software perspective, this simulator is capable of implementing various driving scenarios, including urban, rural, and highway driving. The Pride vehicle's 14° of freedom dynamic model in this simulator enables realistic and real-time implementation of vehicle behaviors.

Fig. 3. Nasir Driving Simulator.

Three vehicles driving in the same direction are assumed to be present in all devised scenarios, as shown in Fig. 4. The front vehicle is in the same lane as the human driver and the rear car is considered in the left lane. The scenario is initiated as the human driver starts the vehicle and increases the speed; other cars also move at a constant distance and speed equal to the human driver. When the speed of the driver reaches 50, 70, and 90 km/h, the front vehicle brakes at an acceleration of −0.7g, equal to −6.867 m/s^2 [28].

Rear Car

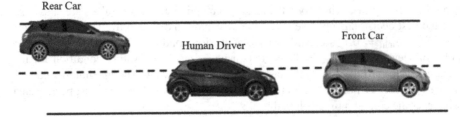

Fig. 4. Vehicle Configurations.

Figure 5, specifies the details regarding the distances in each scenario; two quantities of velocity and headway time are used to determine the distance, d. For each speed, three headway times of 0.5, 1, and 1.5 s were considered, the distance is obtained using these two parameters in each scenario. Also, changes in the headway time parameter directly affect the difficulty level in each section. Eventually, nine different scenarios with different difficulty levels were devised. The parametric values of each scenario and specifications are presented in Table 1.

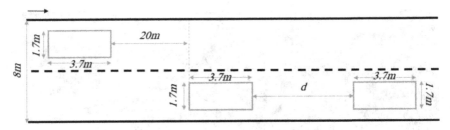

Fig. 5. Scenario description.

Repetitive behavioral changes in both front and rear cars would cause predictability. Therefore, as the headway time increases, three different sudden moves are considered for the rear vehicle so that the human driver cannot avoid the collision by applying the same compensatory behavior. In other words, the repetition could significantly affect the hazard perception assessment, reducing the likelihood of observing the behavioral anomalies in an actual situation. At a headway time of 0.5 s, front and rear cars behave the same when they suddenly break. In scenarios with a headway time of 1 s, the front vehicle suddenly brakes and stops the car; then the rear vehicle also stops with a delay time of 1–2 s after the sudden stop of the front car. In the third scenario, the headway time is 1.5 s, in this scenario, unlike the other two, when the front vehicle stops, the rear car continues on its path with a high speed to hinder decision-making for the human driver.

Table 1. Scenario design parameters.

Rear Car Behavior	d (m)	H (s)	V (km/h)
Sudden Brake	6.549	0.5	50
Sudden Brake with 1–2 s delay	13.89	1	
Continuing With High Speed	20.84	1.5	
Sudden Brake	9.725	0.5	70
Sudden Brake with 1–2 s delay	19.45	1	
Continuing With High Speed	29.17	1.5	
Sudden Brake	12.5	0.5	90
Sudden Brake with 1–2 s delay	25	1	
Continuing With High Speed	37.5	1.5	

6 Hazard Perception Assessment and Behavioral Abnormalities

The study involved a group of ten people between the ages of 22 and 29, with an average age of 25. All participants had a valid driver's license and less than three years of driving experience. Each driver repeated the nine scenarios five times in randomized order to investigate the behavioral changes after repeating each scenario. Driver performance for the first and fifth iterations are listed in Table 2.

Based on the results obtained from each experiment, the statistical values of the driver's initial delay are summarized in Table 3. The order of the scenarios is the same as Table 2. The initial delay implies the time elapsed from the scenario start (front car braking) to the first reaction observed by the human driver. Driver reaction could be a common gesture to prevent the collision. These actions may be removing the foot from the throttle, changes in steering wheel, or pedals. The goal is to find the behavioral anomalies, therefore, the motivations of the driver in different traffic conditions in a driving simulator environment are calculated and plotted.

6.1 Perceptual Anomalies in Reaction Time

Figure 6.a, illustrates the vehicle route in two different cases. The human driver path (dashed line) is compared with the hazard perception delay function (solid line); the delay time is considered equal to the human driver in the same case. It is observed that the delayed model also collides with the front car, implying that the delay of the human driver in the hazard perception is the main cause of the accident. Generally, since there is enough space to change the lanes and the rear car has an adequate distance to the human driver, there will not be a decision-making problem, which means the driver has not shown enough attention in pursuing the front car. Figure 6.b, compares the performance of the human driver (dashed line) with the driver-assist hazard perception function (solid line). The hazard perception function is allowed to interfere with the activation function only when the danger is detected. It is revealed that the model avoids the collision by changing the direction earlier, compared to the human driver.

Table 2. Participants success statistics in each scenario.

Scenario Num	Collision Percentage in All Attempts	Num. of Successful Drivers in The First Attempt	Num. of Successful Drivers in The Fifth Attempt	Rear Car Behavior	H (s)	V (km/h)
8	47	3	10	Brake	0.5	50
6	51	2	5	Brake w/ Delay	1	
3	82	0	2	Increasing Speed	1.5	
1	40	1	9	Brake	0.5	70
9	40	4	7	Brake w/ Delay	1	
5	87	0	3	Increasing Speed	1.5	
4	48	2	8	Brake	0.5	90
2	32	5	8	Brake w/ Delay	1	
7	58	1	7	Increasing Speed	1.5	

Table 3. Participants reaction time in each scenario.

Scenario Num	max of τ_d (s)	min of τ_d (s)	mean of τ_d (s)
8	1.249	0.404	0.662
6	1.384	0.337	0.667
3	1.148	0.473	0.746
1	1.145	0.472	0.63
9	1.678	0.439	0.693
5	1.72	0.337	0.853
4	0.877	0.338	0.574
2	0.975	0.469	0.65
7	1.317	0.47	0.749

In Fig. 6.c, the steering angle of the human driver, the delayed driver, and the driver-assist hazard perception function are shown in dashed, dotted and solid lines respectively. As can be seen, the driver-assist model is steering the wheel only 0.15 s before the human driver, which means that the driver is approximately 0.15 s late from the last moment when it has been possible to escape the collision by turning the steer.

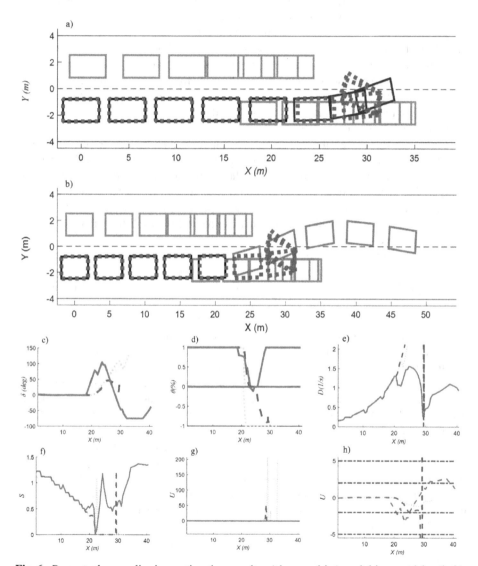

Fig. 6. Perceptual anomalies in reaction time results. a) human driver and driver model path. b) human driver and driver-assist hazard perception model path. c) Steering angle. d) throttle-brake lateral index. e) task demand. f) safety margin. g) affordability. h) lateral slip. Human driver, delayed model and the driver-assist hazard perception model are shown in dashed, dotted and solid lines respectively.

Figure 6.d. is the longitudinal index of using the throttle or brake pedal. Also, it is observed that shortly after steering the wheel, the driver has started to push the brake pedal. The delayed model uses the brake completely (negative values), while the driver-assist model has avoided the collision only by decreasing the pressure on the throttle and steering the wheel in early times, so using the brake has not been a necessity. The delay in decreasing the pressure on the throttle and the rotation of the steering wheel has been equal in both the driver-assist hazard perception model and the human driver.

Figure 6.e. indicates the task demand curve for the human driver, the delayed driver model, and the driver-assist hazard perception model. As shown, the driver-assist hazard perception function has not been activated up to x = 18, and the task demand values associated with the reverse collision time are the same for all three graphs. At x = 20.5m or t = 1.258s, the driver-assist hazard function is activated, and the driver model controls the amount of the task demand. The second hazard perception function is activated when the task demand value exceeds 1.

Figure 6.f. shows the safety margin for the human driver, the delayed driver model, and the driver-assist hazard perception function. It is observed that the driver-assist hazard perception model is activated before the safety margin of the driver reaches zero. On the other hand, the sudden decrease could cause feeling of being threatened. After activating the model with the driver-assist hazard perception function, the safety margin amount is increased again while the human driver has not been able to increase the safety margin, and the collision has occurred.

Figure 6.g. specifies the decision implementation affordability. It can be seen that for the model equipped with the driver-assist hazard perception function, the value is always equal to zero. However, for the delayed driver model and human driver, this value is a positive amount at the moment of collision.

Figure 6.h. illustrates the lateral slip for the human driver, the delayed model, and the driver-assist hazard perception function. It is evident that the human driver has used less steering than the stability limit, due to high perceptual delay and low decision-making time. The model with the driver-assist hazard perception function has inevitably increased the steering to a range between the transition and linear zone. The vehicle has been able to avoid the collision with some lateral slipping but yet a stable movement. The slip angle of the instability threshold is 5 deg, and the saturated slip angle of the linear region is 2 deg.

It is demonstrated that if the initial delay is so large that the driver model cannot avoid the collision with the same delay time, the perceptual defect at the beginning of the scenario could be considered as the cause of the accident, so the initial perception error was detected for this experiment. In the recent experiment, based on the task demand increase, the driver-assist hazard perception function has sensed the danger in time and avoided the collision by controlling the vehicle.

6.2 Decision-Making Anomalies

Figure 7.a. compares the human driver path with the delayed model. As before, the delay is the same as the one for the human driver in the actual test. It can be seen that the driver delay adjusts the delayed model to escape without colliding with the front car, this means that the driver delay is not the cause of the accident. On the other hand, Fig. 7.b.

compares the driver and the driver-assist hazard perception function path, it is indicated that the delay time is not involved because the driver-assist hazard perception function activates the driver model only when the danger is detected. In fact, hazard perception function has been successfully activated after the delay time of the driver.

Figure 7.c. shows the steering angle for the human driver, the delayed driver model, and the driver-assist hazard perception function. As indicated earlier, it can be seen that the driver-assist hazard perception model has turned the steering wheel 0.85 s later than the driver delay time; this means the driver delay has not caused the accident. Also, it can be seen that the model with the driver delay and the model with hazard perception function have both chosen the decision to run away using the steering wheel. In addition, considering the car speed, the longitudinal distance from the front car, the rear car behavior and the available lateral distance, steering is considered to be the most appropriate decision, so the driver has made a mistake in choosing the decision.

Figure 7.d. shows the longitudinal index equal to the throttle and brake pedal usage. It is indicated that the driver applied the brake without turning the steering wheel. The driver model with the delay time pushed the throttle after a short brake (negative values) and uses the steering wheel to escape. The model with the driver-assist hazard perception function also activated after 0.544 s. The model used the brakes harder than the driver and then the steering to avoid the collision; finally, the speed was increased using the throttle.

Figure 7.e. depicts the task demand curve for the human driver, the delayed driver model, and the driver-assist hazard perception model. As before, by increasing the amount of the task demand and passing the value of 1 at x = 29.24 m or t = 1.462 s, the driver-assist hazard function activated, and the task demand was taken into control. Also, it can be seen that the driver overstated the task demand by braking, while both the delayed model and the model with hazard perception function decreased the task demand by applying the steering wheel; so as mentioned before, the driver chose the wrong decision.

Figure 7.f. shows the safe margin diagram for the human driver, the delayed model, and the driver-assist hazard perception function model. After the activation of the hazard perception function and choosing the right decision, the safety margin value increased in contrast to the decision made by the driver. The results are similar for the delayed model behavior, while the human driver decision was unable to increase the safety margin value and collided.

As shown in Fig. 7.h. the lateral slip for the human driver, the delayed model, and the driver-assist hazard perception function, the human driver has stayed at the stable zone without using the steering wheel, due to wrong decision and not trying to compensate it. The model with the hazard perception function inevitably increased the steering to the transition and linear zone. Eventually, the vehicle avoided the collision with some lateral slippage, but with a stable movement. However, considering the appropriate distance, enough time, and less steering, the delayed model used less steering in comparison to the model with the hazard perception function and therefore was more stable.

This section indicated the driver with a decision-making anomaly. Due to the distance between the vehicles and the behavior of the rear vehicle, using the proper delay time, the driver decreased the task demand and safety margin to zero. However, the model

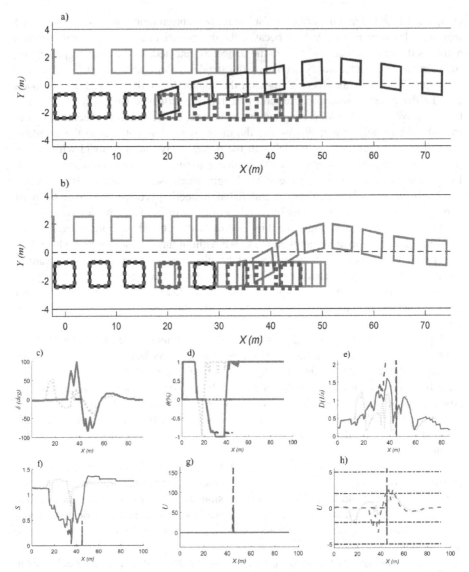

Fig. 7. Decision-making anomalies results. a) human driver and driver model path. b) human driver and driver-assist hazard perception model path. c) Steering angle. d) throttle-brake lateral index. e) task demand. f) safety margin. g) affordability. h) lateral slip. Human driver, delayed model and the driver-assist hazard perception model are shown in dashed, dotted and solid lines respectively.

with the hazard perception function and the model with the delay time made the proper decision and successfully reduced the task demand and prevented decreasing of the safety margin by steering the wheel. Therefore, the diagnosis of the hazard perception error on decision making was experimented in this test.

6.3 Decision-Implementation Anomalies (Low-Gain)

The human driver and the delayed model vehicle path are shown in Fig. 8.a. The obtained results revealed that even in this case, the delay time did not play a role in the collision event; because the delayed model successfully escaped without a collision. Figure 8.b. also confirms the latest conclusion; even though driver-assist hazard perception function activated later, the collision did not occur.

Figure 8.c. shows the steering angle for the human driver, the delayed driver model, and the driver-assist hazard perception function. It can be seen that the driver reacted 0.34 s earlier than the activation of the hazard perception function, therefore, it is assumed that the delay of the driver was not the cause of the accident. Unlike previous cases, the driver decided to use the steering, the same as the model, however, the collision occurred, so, the collision was probably caused by the decision implementation. As mentioned, the driver implements the decision using a controller. An inappropriate gain could beneficially influence the decision implementation. Among available scenarios, the chosen option is one of the simplest. At the final moment, there is an appropriate longitudinal distance from the front car. Also, due to the simultaneous stopping of both rear and front cars, the lateral distance is ideal for escape. As the danger is identified, the driver chooses the right decision and takes action by steering the wheel, however, as shown in Fig. 8.c, the steering angle is less than the delayed model and the model with the hazard perception function; the maximum steering angle is 28.59°, while the delayed model rotates the wheel about 73.58°. The driver selects the best option using the data received from the traffic layout, however, the problem occurs due to incorrect estimation of task demand and capability. In an easy traffic arrangement, the driver overestimates his ability to meet the situation demand, therefore, the driver makes an underestimated decision, considering a low control gain and, accordingly, implementing a slight steering angle to avoid the collision. The wrong estimations could affect other components. Eventually, the driver overlooks the excessive proximity to the front car, and the collision occurs.

Figure 8.d. shows the throttle and brake pedal usage. The driver will use the throttle and brake pedals while steering the wheel. After pushing the brake (negative values) and steering the wheel for a short time, the delayed driver model uses the throttle. Due to time and distance shortage, the model with hazard perception function used a larger steering angle and less braking intensity. After passing the obstacle, the model increased the speed.

The task demand curve for the human driver, delayed driver model, and driver-assist hazard perception function is shown in Fig. 8.e. When the task demand value passes 1, the model with the driver-assist hazard perception function activates and controls the task demand value. The driver had a wrong control gain and incorrect estimation of task demand, therefore, he/she could not comprehend the increased value of task demand and the extreme danger value. In other words, the driver did not sense the danger correctly.

The decision unaffordability is shown in Fig. 8.g. It is observed that for the model equipped with the hazard perception function and the delayed model, this value is always equal to zero. It should be noted that this value was equal to zero for the human driver before the collision occurred, that is, the driver successfully chose the right decision, and he was able to avoid the collision with a slight change in the steering angle even in the last

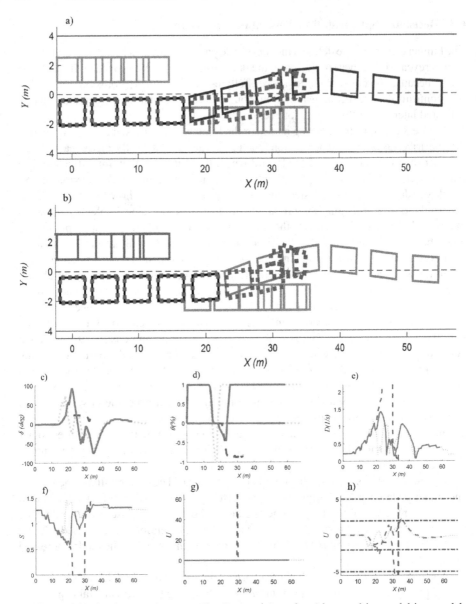

Fig. 8. Decision-implementation anomalies (Low-gain) results. a) human driver and driver model path. b) human driver and driver-assist hazard perception model path. c) Steering angle. d) throttle-brake lateral index. e) task demand. f) safety margin. g) affordability. h) lateral slip. Human driver, delayed model and the driver-assist hazard perception model are shown in dashed, dotted and solid lines respectively.

moments. However, the perform ability is lost due to improper decision implementation, and the collision took place.

In this section, the performance of a driver with a decision implementation anomaly is demonstrated. It was found that the driver has made the right decision due to vehicle distances and the behavior of the rear car. However, the accident has occurred according to implementation anomaly, mainly due to the errors in estimation and inappropriate control gain. It is observed that the perceptual error was detected in the implementation and low-control gain.

6.4 Decision-Implementation Anomalies (High-Gain)

In Fig. 9.a, compares the human driver path and the delayed model. It is evident that the delay time does not play a role in the collision events because the delayed model escaped without collision. Also, looking at Fig. 9.b, the driver-assist hazard perception function activated later and yet could avoid the collision.

Figure 9.c. illustrates the steering angle for the human driver, the delayed driver model, and the driver-assist hazard perception function. In this case, the driver successfully decided, however, the collision occurred. As in the previous case, the collision cause is the decision implementation. Based on the traffic arrangement data, the driver decided and estimated the determinative behavioral factors by selecting the controlling gain for decision implementation. On the other hand, in the previous case, the driver chose a low control gain, leading to a decision implementation anomaly. The present scenario is more complicated than the previous one. However, the steering wheel could be used to escape the situation. After the hazard perception, the driver decided correctly and took action by steering the wheel, however, as shown in Fig. 9.c, the steering angle is more than both the delayed model and the model with the hazard perception function. The human driver escape steering angle is at a maximum of 43° while the model with the hazard perception rotates the steering wheel about 27.62°. In the current situation the speed is 90 km/h, rotating the wheel in such speed can make the car unstable. The driver is in a relatively difficult arrangement and experiences excessive fear and panic, so he/she engages in exaggerated behavior. Therefore, when making the decision, with an exaggerated behavior, the driver considers a high control gain. This exaggerated behavior and estimation miscalculations have affected the components so that the driver can no longer control the car and overreact.

Figure 9.d. shows the longitudinal index equivalent to the driver throttle and brake pedal usage. It is observed that the sense of fear has dramatically impacted the hazard perception, after the beginning of the scenario. Although the driver lost control of the car, he keeps pushing the throttle; but due to slippage, the model with the hazard perception function uses the steering wheel and brakes in the high speed to reduce the throttle intensity.

Figure 9.g. demonstrates the decision unaffordability value. Both the delayed model and the model with the hazard perception function adjusted this amount equal to zero. The human driver also regulated this value close to zero until moments before the accident, which means the driver was able to make the right decision and to confront fear until the last moments, however due to errors in decision implementation, the affordability was lost, and the collision took place.

Figure 9.h. depicts the driver's cause of excessive fear. In the present scenario, due to the high speed of the vehicles, the driver is afraid of slipping and thus chooses a high

116   M. Pashaee and A. Nahvi

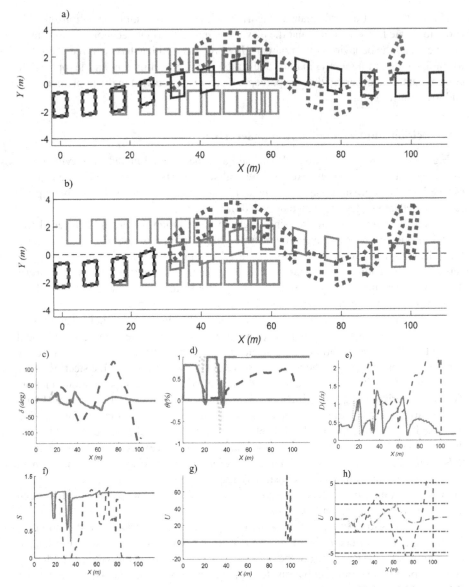

Fig. 9. Decision-implementation anomalies (High-gain) results. a) human driver and driver model path. b) human driver and driver-assist hazard perception model path. c) Steering angle. d) throttle-brake lateral index. e) task demand. f) safety margin. g) affordability. h) lateral slip. Human driver, delayed model and the driver-assist hazard perception model are shown in dashed, dotted and solid lines respectively.

control gain which will lead to car instability. On the other hand, the model with the hazard perception function escaped the situation without getting close to the linear and transition zone; the expression is the same for the delayed model.

In this section, the driver behavior with a decision implementation anomaly is demonstrated. It was found that, considering the distance between the cars and the behavior of the rear car the driver made the right decision. However, due to incorrect estimation and selection of control gain, an improper decision implementation anomaly was encountered. Unlike previous cases, the selected gain is high, and the decision is exaggerated. The present experiment investigated the demonstration of high gain hazard perception for the decision implementation.

7 Conclusion

The present study focuses on utilizing driving motivations as a crucial criterion for analyzing driver behavior. By leveraging physical variables such as time-tocollision, collision avoidance, and execution time, a quantitative assessment of motivations was developed. These motivations include task difficulty, driver capability, safety margin, unaffordability, and fear of slipping. Investigation of these motivational parameters lays the groundwork for developing advanced algorithms to identify driver behavioral anomalies.

To evaluate the algorithm's effectiveness, a group of young drivers underwent comprehensive assessments in a driving simulator encompassing three key categories: reaction, decision-making, and decision execution. The drivers' overall hazard perception skills were also evaluated in various traffic conditions. The results demonstrated the algorithm's ability to enhance drivers' hazard perception skills in a controlled and risk-free simulator environment.

Incorporating hazard perception assessment within the certification process will ensure that only competent drivers with adequate hazard perception skills are granted driving licenses, enhancing road safety and promoting responsible driving behavior among individuals. By emphasizing the importance of hazard perception skills, the training and certification process would play a vital role in creating a safety-aware culture on the roads.

Therefore, the evaluation procedure for hazard perception skills should move beyond traditional methods and embrace more realistic and dynamic approaches to simulate actual driving scenarios. This shift towards practical assessment methods allows for a more accurate and effective evaluation of drivers' hazard perception abilities, leading to improved road safety outcomes.

References

1. Global Status Report on Road Safety - Time for Action | WHO | Regional Office for Africa. https://www.afro.who.int/publications/global-status-report-road-safety-time-action
2. Reason, J., Manstead, A., Stradling, S., Baxter, J., Campbell, K.: Errors and violations on the roads: a real distinction? Ergonomics **33**, 1315–1332 (1990). https://doi.org/10.1080/001401 39008925335
3. History of road safety, The Highway Code and the driving test - GOV.UK. https://www.gov. uk/government/publications/history-of-road-safety-and-the-driving-test/history-of-road-saf ety-the-highway-code-and-the-driving-test

4. Sagberg, F., Bjørnskau, T.: Hazard perception and driving experience among novice drivers. Accid. Anal. Prev. **38**, 407–414 (2006). https://doi.org/10.1016/j.aap.2005.10.014
5. G.B., Groeger, J.A.: Risk, hazard perception, and precieved control. Novice Driveers Conf. (2000)
6. Machin, M.A., Sankey, K.S.: Relationships between young drivers' personality characteristics, risk perceptions, and driving behaviour. Accid. Anal. Prev. **40**, 541–547 (2008). https://doi.org/10.1016/J.AAP.2007.08.010
7. Borowsky, A., Oron-Gilad, T., Meir, A., Parmet, Y.: Drivers' perception of vulnerable road users: a hazard perception approach. Accid. Anal. Prev. **44**, 160–166 (2012). https://doi.org/10.1016/J.AAP.2010.11.029
8. Vlakveld, W.P.: A comparative study of two desktop hazard perception tasks suitable for mass testing in which scores are not based on response latencies. Transp. Res. Part F: Traffic Psychol. Behav. **22**, 218–231 (2014). https://doi.org/10.1016/j.trf.2013.12.013
9. Alavi, S.S., Mohammadi, M.R., Souri, H., Kalhory, S.M., Jannatifard, F., Sepahbodi, G.: Personality, driving behavior and mental disorders factors as predictors of road traffic accidents based on logistic regression. Iran J. Med. Sci. (2017)
10. Horswill, M.S., Hill, A., Buckley, L., Kieseker, G., Elrose, F.: Further down the road: the enduring effect of an online training course on novice drivers' hazard perception skill. Transp. Res. Part F Traffic Psychol. Behav. (2023). https://doi.org/10.1016/j.trf.2023.02.011
11. Äbele, L., Haustein, S., Martinussen, L.M., Møller, M.: Improving drivers' hazard perception in pedestrian-related situations based on a short simulator-based intervention. Transp. Res. Part F Traffic Psychol. Behav. **62**, 1–10 (2019). https://doi.org/10.1016/j.trf.2018.12.013
12. Malone, S., Brünken, R.: The role of ecological validity in hazard perception assessment. Transp. Res. Part F Traffic Psychol. Behav. **40**, 91–103 (2016). https://doi.org/10.1016/j.trf.2016.04.008
13. Markkula, G., Engström, J., Lodin, J., Bärgman, J., Victor, T.: A farewell to brake reaction times? Kinematics-dependent brake response in naturalistic rear-end emergencies. Accid. Anal. Prev. **95**, 209–226 (2016). https://doi.org/10.1016/j.aap.2016.07.007
14. Asadamraji, M., Saffarzadeh, M., Ross, V., Borujerdian, A., Ferdosi, T., Sheikholeslami, S.: A novel driver hazard perception sensitivity model based on drivers' characteristics: a simulator study. Traffic Inj. Prev. **20**, 492–497 (2019). https://doi.org/10.1080/15389588.2019.1607971
15. Sheppard, E., Ropar, D., Underwood, G., Van Loon, E.: Brief report: driving hazard perception in autism. J. Autism Dev. Disord. **40**, 504–508 (2010). https://doi.org/10.1007/s10803-009-0890-5
16. Design, V.: Cognitive and psychomotor correlates of hazard perception ability and risky driving, 211–217 (2000)
17. Gonzalez, D., Perez, J., Milanes, V., Nashashibi, F.: A review of motion planning techniques for automated vehicles. IEEE Trans. Intell. Transp. Syst. **17**, 1135–1145 (2016). https://doi.org/10.1109/TITS.2015.2498841
18. Xue, Q., Ren, X., Zheng, C., Zhou, W.: Research of driver decision making behavior modeling and simulation. In: Proceedings of the 2nd International Conference on Computer Application and System Modeling. Atlantis Press, Paris, France (2012)
19. Wilde, G.J.S.: The theory of risk homeostasis: Implications for safety and health. Risk Anal. **2**, 209–225 (1982). https://doi.org/10.1111/j.1539-6924.1982.tb01384.x
20. Fuller, R.: Towards a general theory of driver behaviour. Accid. Anal. Prev. **37**, 461–472 (2005). https://doi.org/10.1016/j.aap.2004.11.003
21. Gordon, T., Gao, Y.: A flexible hierarchical control method for optimal collision avoidance. In: Proceedings of the 16th International Conference on Mechatronics - Mechatronika 2014, pp. 318–324. IEEE (2014)

22. Hayashi, R., Isogai, J., Raksincharoensak, P., Nagai, M.: Autonomous collision avoidance system by combined control of steering and braking using geometrically optimised vehicular trajectory. Veh. Syst. Dyn. **50**, 151–168 (2012). https://doi.org/10.1080/00423114.2012.672748

23. Gibson, J.J., Crooks, L.E.: A theoretical field-analysis of automobile-driving. Am. J. Psychol. **51**, 453 (1938). https://doi.org/10.2307/1416145

24. Engström, J., Hollnagel, E.: A general conceptual framework for modelling behavioural effects of driver support functions. In: Modelling Driver Behaviour in Automotive Environments, pp. 61–84. Springer London, London (2007)

25. Hollnagel, E.: Time and time again. Theor. Issues Ergon. Sci. **3**, 143–158 (2002). https://doi.org/10.1080/14639220210124111

26. Mozaffari, E., Nahvi, A.: A motivational driver model for the design of a rear-end crash avoidance system. Proc. Instit. Mech. Eng. Part I: J. Syst. Control Eng. **234**(1), 10–26 (2020). https://doi.org/10.1177/0959651819847380

27. Gáspár, P., Szabó, Z., Bokor, J., Nemeth, B.: Robust Control Design For Active Driver Assistance Systems. Springer International Publishing, Cham (2017). https://doi.org/10.1007/978-3-319-46126-7

28. Lee, S.E., Llaneras, E., Klauer, S.G., Sudweeks, J.: Analyses of rear-end crashes and near-crashes in the 100-car naturalistic driving study to support rear-signaling countermeasure development. Distribution (2007)

Driver Identification by an Ensemble of CNNs Obtained from Majority-Voting Model Selection

Rouhollah Ahmadian[1] , Mehdi Ghatee[1(✉)] , and Johan Wahlström[2]

[1] Department of Computer Science, Amirkabir University of Technology, Hafez Ave., Tehran 15875-4413, Iran
{rahmadian,ghatee}@aut.ac.ir
[2] Department of Computer Science, University of Exeter, Exeter EX4 4QF, UK
J.Wahlstrom@exeter.ac.uk

Abstract. Driver identification refers to the task of identifying the driver behind the wheel among a set of drivers. It is applicable in intelligent insurance, public transportation control systems, and the rental car business. An critical issue of these systems is the level of privacy, which encourages a lot of research using non-visual data. This paper proposes a novel method based on IMU sensors' data of smartphones. Also, an ensemble of convolutional neural networks (CNNs) is applied to classify drivers. Furthermore, the final prediction is obtained by a majority vote mechanism. This paper demonstrates that model selection using a majority vote significantly improves the accuracy of the model. Finally, the performance of this research in terms of the accuracy, precision, recall, and f1-measure are 93.22%, 95.61%, 93.22%, and 92.80% respectively when the input length is 5 min.

Keywords: Driver Identification · Deep Learning · Majority Vote · IMU Sensors · Smartphone Data

1 Introduction

This study aims to identify the driver from a set of drivers based on driving styles [10]. Various sensors are used for this problem; see Table 1 for a review of related research. Researchers often use CAN-bus technology to utilize information from the Engine Control Unit (ECU), like engine speed, brake pedal, fuel consumption, and steering angle signals. However, the cost of CAN-bus technology is high, and not all vehicles support this technology. Instead of ECU data, the Inertial Measurement Unit (IMU) can be the right choice, which includes accelerometers, gyroscopes, and magnetometers. Accelerometers measure acceleration, gyroscopes measure angular velocity, and magnetometers measure the magnetic field. Since all smartphones contain IMU sensors, driving identification can be performed using smartphone data. In driver identification using machine learning methods, the following parameters should be considered:

– The sliding window length,

M. Ghatee and S. M. Hashemi (Eds.): ICAISV 2023, CCIS 1883, pp. 120–136, 2023.
https://doi.org/10.1007/978-3-031-43763-2_8

- The percentage of overlap between sequential windows,
- The amount of training data,
- And the number of drivers.

There have been numerous studies conducted on driver identification using deep learning methods. For instance, in [9], turning maneuver data is utilized by clustering 12 ECU signals into 12 sets. A random forest classifier is then used to classify drivers based on these sets. However, this research's weakness is that it only focuses on a specific maneuver. In [17], eight ECU signals are used to classify drivers using an extreme learning machine. The extraction method of features is crucial in all of these attempts, with recent works using both explicit and implicit features. The explicit approach involves extracting statistical, spectral, and temporal features. On the other hand, the implicit approach extracts hidden features using deep networks. [24] describes a method for identifying drivers using a histogram of acceleration data as input to a neural network. [14] extracts statistical features, including mean, max, min, median, standard deviation, skew, kurtosis, 25% quartiles, and 75% quartiles. In [2], statistical, spectral, and temporal features of accelerometer and gyroscope data are compared. Also, it was proved that the histogram is the best feature for driver identification. In addition, a Generative Adversarial Network is used to augment training data. [22] proposes a driver identification system based on the hidden features of acceleration data. The longitudinal and transverse signals of the accelerometer and angular velocity are transmitted to the two-dimensional space using a two-layer convolutional network. The data are then classified using the ResNet-50 + GRU network. In this method, features are extracted by convolutional layers. Further, [25] proposes a technique based on a fully convolutional network and squeeze-and-excitation (SE) blocks. The method collects psychological data from a driving simulator, including vehicle control operation and eye movement data. Additionally, a method based on incremental learning is proposed in [26] for scalable driver authentication in dynamic scenarios, using SVMs and convolutional neural networks (CNNs) in the model.

In assessing the performance of a driver identification model, the following scientific gaps can be identified in previous research:

- Unlike the problem of driving evaluation, which only requires one model that works for every driver, identifying a driver among a set of drivers requires individual models for every driver set due to the nature of identification. However, from a business perspective, because collecting such data can be costly, approaching with minor training data in high demand is essential. Section 3.5 demonstrates that our model requires less training data than previous works.
- High-performance driver identification requires quick identification. For example, in the case of car theft, an alarm that goes off after an hour is not helpful. In other words, the model should quickly identify the driver with a small amount of data, with significant overlap between the input windows.
- In addition, the model's accuracy should remain relatively high as the number of drivers in the set increases.

We have designed a paradigm of sensitivity analyses to address the gaps in previous research. Our previous study examined a wide range of obvious features, including statistical, spectral, and temporal features, and found that identifying the most useful features in this feature space is difficult and time-consuming. Therefore, our proposed model is based on extracting hidden patterns from driving data using deep learning, which achieves greater accuracy than previous works that use traditional feature extraction methods. However, the CNN model we used experienced overfitting and instability. To overcome these issues, we introduced a model selection method based on majority voting to select the best model.

Our contributions can be summarized as follows:

1. The introduction of a standard template for benchmarking driving identification models.
2. Use a majority-voting mechanism as a model selection to overcome overfitting.

In this study, we used data collected by [5] to train and evaluate our model. This dataset was collected in sunny weather conditions in Tehran city and included data from ten drivers in various vehicles. The drivers were in a normal mental state without any stress, and approximately 7 h and 45 min of data were available for each driver. The dataset includes an accelerometer, gyroscope, magnetometer field, thermometer, and GPS data. However, we only used the acceleration, angular velocity, and magnetometer field in three spatial directions for this study. The data sampling rate for inertial sensors was 2 Hz, and the GPS sensor was sampled at 1 Hz.

This study uses the data collected by [5] to train and evaluate the model. This dataset was collected in sunny weather conditions in Tehran city and included data from ten drivers in various vehicles. The drivers were in a normal mental state without any stress, and approximately 7 h and 45 min of data were available for each driver. The data set includes the accelerometer, gyroscope, magnetometer field, thermometer, and GPS data. As mentioned before, this study uses only the acceleration, angular velocity, and magnetometer field in three spatial directions. The data sampling rate was 2 Hz and 1 Hz for the inertial sensors and the GPS receiver, respectively.

Table 1. Selected driver identification studies [2].

Ref	Dataset	Sources	Signals	Features	Models	Drivers	Accuracy
[6]	Private owner	CAN-bus	48 ECU signals	Statistical, Descriptive, Frequency	SVM, RF, Naive Bayes, KNN	15	100%
[17]	[1]	CAN-bus, Smartphone	12 ECU + 6 IMU + headway distance	Temporal, Frequency Domain, Cepstral	ELM	11	84.36%
[9]	Private owner	CAN-bus	12 ECU signals	Statistical, Spectral, PCA, DWT	Random Forest	2	76.9%
						3	59.4%
						4	55.2%
						5	50.1%
[8]	[15]	CAN-bus & GPS	Path, Speed, Jerk, Duration, Acceleration	Statistical	Linear Discriminant Analysis (LDA)	5	60.5%
[11]	[1]	CAN-bus	5 ECU signals	Statistical, Cepstral	SVM, RF, AdaBoost, ET	5	95%
						15	89%
						35	82%
[12]	[1]	CAN-bus	2 ECU signals	Spectral, Cepstral	Gaussian Mixture Models	5	99.99%
						15	99.7%
						35	99.6%
						50	99.5%
						67	99.5%
[16]	[13]	CAN-bus	6 ECU signals	PCA	Decision Tree	10	99%
[7]	[13]	CAN-bus	51 ECU signals	Statistical	Extra Trees	10	99.92%
	[21]	GPS	8 GPS signals			6	76%
	[23]	Temperature, Webcam, GPS	Brightness, Acceleration, GPS, Physiological Signals			10	100%
[3]	Private owner	GPS	Speed, Heading	Random Forest	Statistical	38	82.3%
[20]	Private owner	Smartphone	Acceleration	Statistical	PCA	5	60%-100%
[24]	Private owner	Raspberry pi, GPS	Acceleration	Histogram	MLP	13	94.77%
[22]	Private owner	Smartphone	Acceleration, Angular Velocity	CNN feature maps	Resnet50 + GRU	25	71.89%
[27]	Private owner	Event Simulator	Acceleration, Brake, Steering Wheel	Driving manoeuvres such as deceleration, change lane	Hidden Markov Model	20	85%
[18]	Private owner	IVDR	Location, manoeuvres	Statistical	Stacked generalization method	217 families	88%
[19]	Private owner	CAN-bus + CarbigsP	51 ECU features	Exclusion highly correlated features	GAN's discriminator	4	88%

2   Proposed System

The raw driving data comes from acceleration and gyroscope sensors. Driving data consists of driving maneuvers such as lane change, deceleration, acceleration, left and right turns, and stops. Each driver often has unique maneuvers because of different driving styles. However, while each driver has a different driving style, the drivers drive similarly in many cases. The drivers are only distinct with respect to specific events. For example, one driver brakes uniquely, and another turns around differently, but both change lines in the same way. In other words, a significant portion of the driver data is similar, which is an issue for driver identification. This issue means that only some pieces of data are suitable for identifying the driver. Some research has considered high-dimensional input data to overcome this problem. Nevertheless, while increasing the input data size increases the likelihood of retrieving the appropriate features, the curse of high dimensionality also arises. This paper identifies drivers by extracting hidden patterns from acceleration and gyroscope signals using a deep Convolutional Neural Network (CNN). Furthermore, the majority vote solves the issue of long input while keeping the classifier's input size small. The experimental results show that the model's accuracy is significantly increased. Figure 1 shows a novel architecture for driver identification. The proposed architecture includes two modules of pre-processing and identification. In the following, these modules are described in detail.

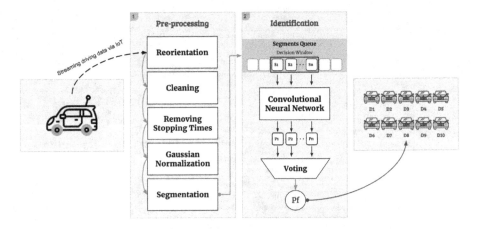

Fig. 1. The architecture of a driver identification system.

2.1 Data Pre-processing

In this module, acceleration, angular velocity, and magnetic field signals are preprocessed in six steps, including axis alignment, cleaning, removal of motionless periods, split into test and training sets, Gaussian normalization, and segmentation. The magnetic signal is used only in the alignment step. The input data contains three signals, and each signal includes three axes, so the output contains nine segmented signals. Table 2 shows the available driving data for analysis.

Reorientation. In this step, the orientation of the device used for data collection is aligned with the vehicle axes using acceleration and magnetic field data [4].

Cleaning. First, outliers in the signals are clipped based on the maximum and minimum values of non-outliers data points, and then the missing values are replaced with the average of each signal.

Table 2. The amount of remaining driving data after removing motionless periods

Driver Id	Before	After	Driver Id	Before	After
201	04:19:09	01:40:28	206	06:44:40	01:57:58
202	06:53:36	01:54:35	207	12:13:19	03:02:44
203	08:24:28	02:43:11	208	08:00:58	02:06:49
204	08:39:25	01:49:25	209	06:34:13	02:13:29
205	09:38:07	01:27:43	210	06:05:37	02:50:59

Motionless Removing. When the vehicle is at standstill for a long period of time, the data is not useful in analyzing the driver's behavior. Therefore, we discard a data point if the vehicle has stopped for more than 6 s. The sum of 6 s acceleration norm-2 is calculated, then if it is below a predefined threshold θ, the vehicle is assumed immobile, see Eq. 1.

$$\sum_{t}^{t+6*f_s} \sqrt{a_{x,t}^2 + a_{y,t}^2 + a_{z,t}^2} \leq \theta \tag{1}$$

where the threshold θ is 0.5, and the frequency f_s is 2 Hz.

Data Splitting. During the model training phase, the data is divided into training and testing data. Although the training data size is not consistent in various experiments, the testing data size is always 30 min.

Gaussian Normalization. The data is normalized with the training data configuration by the Gaussian formula 2. As the six signals' modes are different, their standardization does not affect the correlation between these signals. Statistics show that they cause the network to converge.

$$x' = \frac{x - \bar{x}}{\sigma} \tag{2}$$

Segmentation. In the last, each signal is split into several segments using a sliding window. The number of segments is a function of the length and overlap of the sliding windows. It is essential to note that training and testing windows in the training model phase have nothing in common.

2.2 Identification

This module aims to classify the drivers using hidden pattern extraction from segmented data. The proposed model consists of two stages. First, a CNN model classifies segments inside the decision window. Finally, the most repetitive prediction is selected. The model first learns from training data and is then evaluated with testing data.

Classification Model. A wide range of CNN architectures is evaluated to classify drivers in this stage. We investigated some architectures for classifiers based on some arrangement of data in the segmentation step. Since there is no correlation between the axes of sensors, one way to build a classifier is by using multiple CNN and concatenating their feature maps. Therefore, data of each axis of sensors, a vector, is given to a CNN. This approach considers data to be multi-modal and uses a fusion of the hidden features before the fully connected model. On the other hand, the architecture of the classifier can be only one CNN. In this case, there are three ways to arrange the raw data. First, each axis of the sensors can be put together inside a vector. Second, each of them can take a row in a matrix. Third, each of them can take a channel in a tensor. Thus, four general approaches are used for extracting hidden features based on the arrangement of data:

1. **One 1-D CNN with one channel (1-D):** all vectors are concatenated to a vector. The input data shape is $[n, 6*w]$. Where n is the number of segments, and w is the sliding window length.
2. **One 2-D CNN with one channel (2-D):** all vectors are stacked in a matrix. The input data shape is $[n, 1, 6, w]$.
3. **One 1-D CNN with six channels (3-D):** each vector is put in one channel. The input data shape is $[n, 6, 1, w]$.
4. **Six 1-D CNN with one channel (4-D):** there are six CNN connected to an MLP in this case. The input data shape is $[6, n, w]$.

Each approach has different divisions based on the layers' numbers, types, and attributes. Moreover, we test several deep convolutional architectures for each general approach. The experimental results section presents the best architecture of these four primary approaches.

Majority Vote. In this step, we want to boost the accuracy of the model. In the evaluation of the model, false negative predictions are high because the sliding window length is small in the segmentation step. Instead of predicting a small part of driving, we predict the driver using a time period that is longer than the original segmentation window. This longer time period is referred to as the decision window. There are n segments in each decision window. This means that the decision window length is significantly larger than the segments' length. For example, if the segmentation window length is 4 s, and the decision window length could be 10 min. Also, decision windows have an overlap. Finally, the class with the most repetitions in the prediction set is selected as the final prediction. Thereby, n predictions are reduced to 1 while increasing the chance of correct classification. Figure 2 shows the architecture of predicting by majority vote.

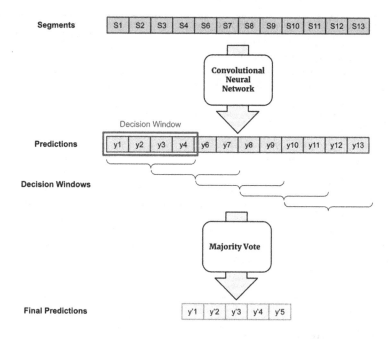

Fig. 2. The architecture of predicting by majority vote.

2.3   Network Training Method

This paper proposes a new method to apply the influence of the majority vote on model selection based on network loss. Early stopping is a regularization method for controlling overfitting. The early stopping evaluates the network with validation data at the end of each epoch in the training process. The training process is stopped when the model has improved sufficiently on the validation set. Therefore, early stopping can not apply to our proposed model because the network's weights are updated based on the loss of objective function. As it is independent of the majority vote, a better local minimum may be obtained by continuing the training.

Our proposed model selection method does not stop before specified epochs. At the end of each epoch, the accuracy based on the majority vote is calculated for the validation set. If the accuracy is maximum, the weight of the network is kept in a temporal container. After training, the weights in the container are replaced. Details of the proposed algorithm can be seen in pseudo-code 1.

Algorithm 1. The proposed method Network by majority vote

Require: S_{train}: the train segments
Require: $S_{validation}$: the validation segments
Require: dt: the decision window length
Require: do: the decision window overlap size
    for $k \leftarrow 1$ to $epochs$ **do**
        $L \leftarrow Loss(\theta, S_{train})$
        $\theta \leftarrow backpropagation(\theta, L)$
        $y_{predicts} \leftarrow predict(\theta, S_{validation})$
        $y'_{predicts}, y'_{truth} \leftarrow Segmentation(y_{predicts}, y_{truth}, dt, do)$ ▷ Obtaining the decision windows
        $\hat{y}_{predicts} \leftarrow MajorityVote(y'_{predicts})$ ▷ Calculating the voting prediction
        $accuracy \leftarrow ControllingMeasure(y'_{truth}, \hat{y}_{predicts})$ ▷ Calculating the final accuracy
        if $accuracy \geq best$ **then**
            $best \leftarrow accuracy$
            $\theta_{best} \leftarrow \theta$ ▷ Holding the best weights
        end if
        $\theta \leftarrow \theta_{best}$
    end for

---

3   Experimental Results

The present section of the manuscript outlines an evaluation of the proposed system. The primary objective of this evaluation is to identify the optimal configuration for the system with respect to its network architecture and associated parameters. Initially, the network architecture shall be subjected to a thorough examination to determine the most effective and efficient design. Subsequently, the optimal sliding window length and overlap, the optimal length and overlap for decision windows, as well as the minimum training data size shall be identified. The identification of these parameters is crucial in enhancing the performance of the proposed system and ensuring its suitability for the intended application.

3.1   Sensitivity Analysis on Learning Models

Numerous CNN architectures for the main approaches could be used with different parameters, types, and attributes of layers. Therefore, several experiments have been conducted in this project. Figure 3 shows the best architecture for each primary approach in terms of running cost and accuracy. The dropout and batch normalization layers helped the model to converge and enhanced the generalization performance. In the following, these three architectures are going to be compared to find the best architecture. Figure 4 presents information about the accuracy of the architectures in Fig. 3 in terms of the majority vote (MV) based on the decision window length. Each point in the figure displays the average accuracy of many experiments so that the decision window in the majority-voting stage has a specific length in the minute scale. Colors separate the different architectures. In all experiments of this stage, ten drivers participate, and also 45 min of each driver's data are used to train the model. The sliding window length and overlap are 75% and 4 s, respectively, in the segmentation step. As the figure shows, the accuracy of the proposed system with 4-D CNN is often higher than others. Furthermore, the highest accuracy in this analysis is 95.4% when the decision window length is 9 min.

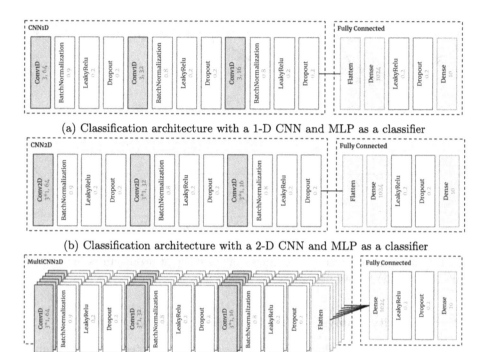

(a) Classification architecture with a 1-D CNN and MLP as a classifier

(b) Classification architecture with a 2-D CNN and MLP as a classifier

(c) Classification architecture with multiple 1-D CNN and MLP as a classifier

Fig. 3. Top convolutional network architectures for driver identification

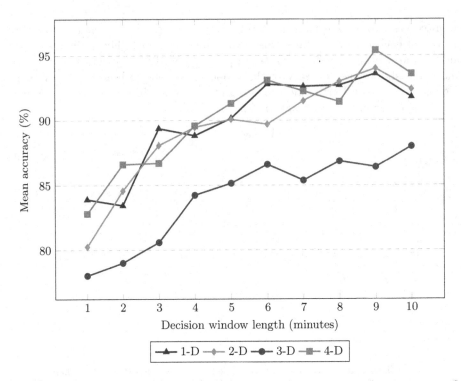

Fig. 4. Comparison between the accuracies of the different architectures in terms of majority vote (MV) based on decision window length.

3.2 Sensitivity Analysis on Segmentation Window Length

In the pre-processing module, the given signals are segmented using the sliding window in the segmentation. We have carried out several experiments using a 4-D CNN architecture as the basis for the experiments. Figure 5 presents information about the accuracy of the proposed model in terms of majority vote (MV) based on the sliding window length in the segmentation stage. Each bar in the figure displays the average accuracy of many experiments so that the sliding window in the segmentation step has a specific length in the second scale. Colors separate the different percentages of sliding window overlap. In all experiments of this stage, ten drivers participate, and also 45 min of each driver's data are used to train the model. The decision window length and its overlap are 75% and 9 min, respectively, in the segmentation step. As can be observed, the best accuracy is 95.4%, and the best-performing sliding window length is 4 s with 75% overlap.

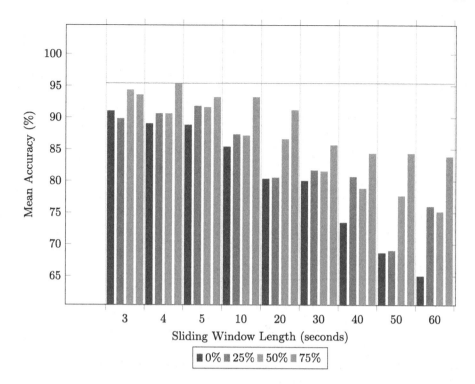

Fig. 5. Comparison of mean accuracy in terms of majority vote (MV) based on sliding window length.

3.3 Sensitivity Analysis on Decision Window Length

This section attempts to find the best length and overlap of the decision window. We evaluated the proposed system based on 4-D CNN architecture with several experiments. In all cases, the sliding window length and overlap were 5 min and 75%, respectively. Figure 6 shows the average accuracy of the proposed model. The model has been evaluated under the decision window lengths from 1 to 15 min and overlap sizes of 0, 25, 50, and 75%. The figure shows that the highest mean accuracy is 95% when the decision window length is 10 min and the overlap size is 25%. However, with small decision windows and a large overlap size, the driver will be identified sooner, and also, there will be a large number of predictions. From this point of view, when the decision window length is 6 min with 75% overlap, the mean accuracy of the proposed system is efficient by 94% majority vote accuracy. Generally, the selection of the optimal decision window length and overlap size depends on the application.

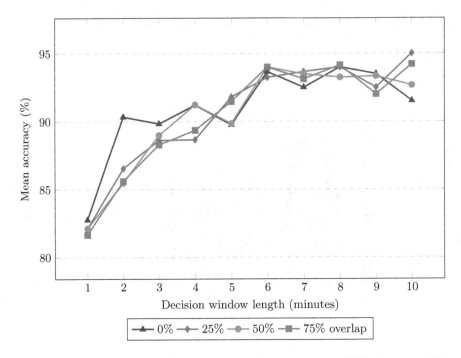

Fig. 6. Comparison of mean accuracy in terms of majority vote (MV) based on sliding window length and overlap percentage.

3.4   Sensitivity Analysis on Training Data Length

This section finds the optimal size for training data in the training phase. We implemented experiments under various driver numbers, including randomly 4, 6, 8, and 10 drivers, based on 4-D CNN architecture. In this section's experiments, the sliding window length and overlap size were always 4 s and 75%, respectively, and the decision window length and overlap were 5 min and 75%, respectively. The training data size ranged between 5 and 45 min. Also, the 30 min of testing data were the same for all experiments. This means that the conditions of evaluation for all models were the same. Figure 7 provides the proposed system's average accuracy in terms of majority vote (MV) based on training data size. As can be seen, the average accuracy is the highest when the training data size is 45 min for each driver. Furthermore, the figure indicates that the variance of the accuracy tends to become smaller when the training data size increases.

4   Performance of the Proposed Model

This section compares this research with state-of-the-art driver identification research based on inertial sensor data. Our proposed system is based on 4-D CNN architecture, using 4 s sliding windows with 75% overlap. In addition, majority

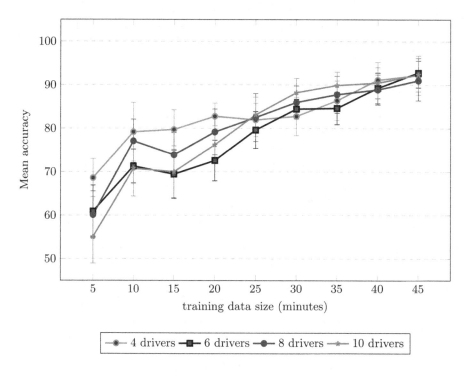

Fig. 7. Comparison of mean accuracy in terms of majority vote (MV) based on training data size.

voting is based on 5 min decision window with a 75% overlap, and also the training data size was 45 min per driver.

The best parameters of our previous research presented in [2] were less efficient in comparison with this paper. The outcomes were based on the 15-minute sliding window, and also the training data size was 90 min per driver. Furthermore, we will also make comparisons with other research. The paper [24] indicates that their training data is 10 h per driver. The sliding window length and overlap sizes are 45 min and 75%. In [14], the training data is around 15 h per driver, and the sliding window length and overlap size are 10 s and 50%. In summary, this paper significantly reduces the decision time and also training data size.

Furthermore, we implemented all these studies according to their papers and the results are available to the public[1]. Therefore, we evaluated some studies with the same parameters based on our data set. Figure 8 compares the result of implementations in terms of average accuracy based on the number of drivers. In this experiment, we used the same parameters for all studies. The sliding window length and its overlap are 4 s and 75%. The decision window length is 5 min with 75% overlap. The training data size for each driver is 45 min. Also, our model

---

[1] https://github.com/Ruhallah93/Driver-Identification.

is a 4-D CNN. As can be seen, our proposed model's accuracy is significantly better than others, and in addition, the accuracy does not fluctuate.

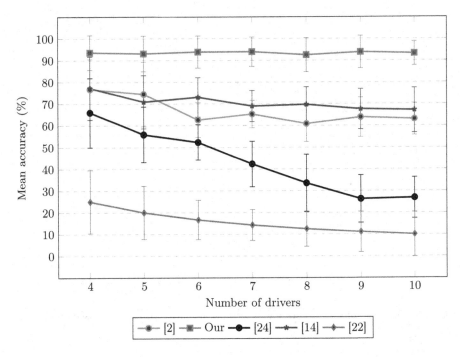

Fig. 8. Comparison between the accuracy of driver identification studies based on the growing number of drivers in terms of majority vote (MV).

5   Conclusion

This paper presents an architecture for driver identification that employs convolutional neural networks to extract hidden features from the data. By applying the majority vote's decision function to a 4-second sliding window with 75% overlap, the accuracy of the convolutional neural network's results is significantly improved. Previous work proposed an optimal system input length of 15 min and 75% overlap, with a training period of 90 min per driver. One of the primary goals of this paper was to reduce the system input length, resulting in an optimal length of 6 min with a 75% overlap. Another objective was to reduce the size of the training dataset, which was decreased to 45 min. The proposed model achieved an accuracy, precision, recall, and f1-measure of 93.22%, 95.61%, 93.22%, and 92.80%, respectively, across ten drivers with a 5-minute decision window length. Compared to our previous work [2], this paper increased the accuracy by about 30%.

References

1. Abut, H., et al.: Real-world data collection with "UYANIK". In: Takeda, K., Erdogan, H., Hansen, J.H.L., Abut, H. (eds.) In-Vehicle Corpus and Signal Processing for Driver Behavior, pp. 23–43. Springer, Boston (2009). https://doi.org/10.1007/978-0-387-79582-9_3
2. Ahmadian, R., Ghatee, M., Wahlström, J.: Discrete wavelet transform for generative adversarial network to identify drivers using gyroscope and accelerometer sensors. IEEE Sens. J. **22**(7), 6879–6886 (2022)
3. Chowdhury, A., Chakravarty, T., Ghose, A., Banerjee, T., Balamuralidhar, P.: Investigations on driver unique identification from smartphone's GPS data alone. J. Adv. Transport. **2018** (2018)
4. Eftekhari, H.R., Ghatee, M.: An inference engine for smartphones to preprocess data and detect stationary and transportation modes. Transport. Res. Part C Emerg. Technolog. **69**, 313–327 (2016)
5. Eftekhari, H.R., Ghatee, M.: Hybrid of discrete wavelet transform and adaptive neuro fuzzy inference system for overall driving behavior recognition. Transport. Res. F Traff. Psychol. Behav. **58**, 782–796 (2018)
6. Enev, M., Takakuwa, A., Koscher, K., Kohno, T.: Automobile driver fingerprinting. Proc. Privacy Enhanc. Technol. **2016**(1), 34–50 (2016)
7. Ezzini, S., Berrada, I., Ghogho, M.: Who is behind the wheel? driver identification and fingerprinting. J. Big Data **5**(1), 9 (2018)
8. Fung, N.C., et al.: Driver identification using vehicle acceleration and deceleration events from naturalistic driving of older drivers. In: 2017 IEEE International Symposium on Medical Measurements and Applications (MeMeA), pp. 33–38. IEEE (2017)
9. Hallac, D., et al.: Driver identification using automobile sensor data from a single turn. In: 2016 IEEE 19th International Conference on Intelligent Transportation Systems (ITSC), pp. 953–958. IEEE (2016)
10. Igarashi, K., Miyajima, C., Itou, K., Takeda, K., Itakura, F., Abut, H.: Biometric identification using driving behavioral signals. In: 2004 IEEE International Conference on Multimedia and Expo (ICME)(IEEE Cat. No. 04TH8763), vol. 1, pp. 65–68. IEEE (2004)
11. Jafarnejad, S., Castignani, G., Engel, T.: Towards a real-time driver identification mechanism based on driving sensing data. In: 2017 IEEE 20th International Conference on Intelligent Transportation Systems (ITSC), pp. 1–7. IEEE (2017)
12. Jafatnejad, S., Castignani, G., Engel, T.: Revisiting gaussian mixture models for driver identification. In: 2018 IEEE International Conference on Vehicular Electronics and Safety (ICVES), pp. 1–7. IEEE (2018)
13. Kwak, B.I., Woo, J., Kim, H.K.: Know your master: driver profiling-based anti-theft method. In: PST 2016 (2016)
14. Li, Z., Zhang, K., Chen, B., Dong, Y., Zhang, L.: Driver identification in intelligent vehicle systems using machine learning algorithms. IET Intel. Transport Syst. **13**(1), 40–47 (2018)
15. Marshall, S.C., et al.: Protocol for candrive ii/ozcandrive, a multicentre prospective older driver cohort study. Accid. Anal. Prevent. **61**, 245–252 (2013)
16. Martinelli, F., Mercaldo, F., Orlando, A., Nardone, V., Santone, A., Sangaiah, A.K.: Human behavior characterization for driving style recognition in vehicle system. Comput. Electr. Eng. **83**, 102504 (2020)

17. Martínez, M., Echanobe, J., del Campo, I.: Driver identification and impostor detection based on driving behavior signals. In: 2016 IEEE 19th International Conference on Intelligent Transportation Systems (ITSC), pp. 372–378. IEEE (2016)
18. Moreira-Matias, L., Farah, H.: On developing a driver identification methodology using in-vehicle data recorders. IEEE Trans. Intell. Transp. Syst. 18(9), 2387–2396 (2017)
19. Park, K.H., Kim, H.K.: This car is mine!: Automobile theft countermeasure leveraging driver identification with generative adversarial networks. arXiv preprint arXiv:1911.09870 (2019)
20. Phumphuang, P., Wuttidittachotti, P., Saiprasert, C.: Driver identification using variance of the acceleration data. In: 2015 International Computer Science and Engineering Conference (ICSEC), pp. 1–6. IEEE (2015)
21. Romera, E., Bergasa, L.M., Arroyo, R.: Need data for driver behaviour analysis? Presenting the public UAH-driveset. In: 2016 IEEE 19th International Conference on Intelligent Transportation Systems (ITSC), pp. 387–392. IEEE (2016)
22. Sánchez, S.H., Pozo, R.F., Gómez, L.A.H.: Driver identification and verification from smartphone accelerometers using deep neural networks. IEEE Trans. Intell. Transport. Syst. (2020)
23. Schneegass, S., Pfleging, B., Broy, N., Heinrich, F., Schmidt, A.: A data set of real world driving to assess driver workload. In: Proceedings of the 5th International Conference on Automotive User Interfaces and Interactive Vehicular Applications, pp. 150–157 (2013)
24. Virojboonkiate, N., Chanakitkarnchok, A., Vateekul, P., Rojviboonchai, K.: Public transport driver identification system using histogram of acceleration data. J. Adv. Transport. 2019 (2019)
25. Xu, J., Pan, S., Sun, P.Z., Park, S.H., Guo, K.: Human-factors-in-driving-loop: driver identification and verification via a deep learning approach using psychological behavioral data. IEEE Trans. Intell. Transport. Syst. (2022)
26. Xun, Y., Guo, W., Liu, J.: G-driveraut: a growable driver authentication scheme based on incremental learning. IEEE Trans. Veh. Technol. (2023)
27. Zhang, X., Zhao, X., Rong, J.: A study of individual characteristics of driving behavior based on hidden Markov model. Sens. Transducers 167(3), 194 (2014)

State-of-the-Art Analysis of the Performance of the Sensors Utilized in Autonomous Vehicles in Extreme Conditions

Amir Meydani$^{(\boxtimes)}$

Amirkabir University of Technology, Tehran, Iran
Amirm199628@gmail.com

Abstract. Today, self-driving car technology is actively being researched and developed by numerous major automakers. For autonomous vehicles (AVs) to function in the same way that people do—perceiving their surroundings and making rational judgments based on that information—a wide range of sensor technologies must be employed. An essential concern for fully automated driving on any road is the capacity to operate in a wide range of weather conditions. The ability to assess different traffic conditions and maneuver safely provides significant obstacles for the commercialization of automated cars. The creation of a reliable recognition system that can work in inclement weather is another significant obstacle. Unfavorable climate, like precipitation, fog, and sun glint, and metropolitan locations, with their many towering buildings and tunnels, which can cause problems or impair the operation of sensors, as well as many automobiles, pedestrians, traffic lights, and so on, all present difficulties for self-driving vehicles. After providing an overview of AVs and their development, this paper evaluates their usefulness in the real world. After that, the sensors utilized by these automobiles are analyzed thoroughly, and subsequently, the operation of AVs under varying settings is covered. Lastly, the challenges and drawbacks of sensor fusion are described, along with an analysis of sensor performance under varying environmental circumstances.

Keywords: Autonomous Vehicle (AV) · Self-Driving Car · Autonomous Driving (AD) · LiDAR · Sensor Fusion

1 Introduction

The World Health Organization (WHO) reports that in 2018, road traffic accidents claimed the lives of 1.35 million people worldwide, making them the eighth biggest cause of accidental death worldwide. The number of reported yearly road fatalities in the European Union (EU) has decreased, but it still exceeds 40,000, with 90% attributable to human error. Due to this, and because they have the potential to enhance traffic flow, investors around the world have contributed much to the research and development of autonomous cars. Furthermore, The introduction of AVs is expected to contribute to the achievement of carbon emission reduction goals [1]. Automated cars can transport

© The Author(s), under exclusive license to Springer Nature Switzerland AG 2023
M. Ghatee and S. M. Hashemi (Eds.): ICAISV 2023, CCIS 1883, pp. 137–166, 2023.
https://doi.org/10.1007/978-3-031-43763-2_9

people and goods just like regular cars, but they can sense their surroundings and find their way around with hardly any help from a human driver. The global AV market size is estimated to be over 6,500 units in 2019, with a CAGR of 63.5% from 2020 to 2027, as stated in a report by Precedence Research [2]. The concept of driverless cars has been around since at least 1918 [3], and it was even discussed on television as early as 1958. In 1988, a demonstration was given at Carnegie Mellon University showing how the NAVLAB vehicle could follow lanes using video footage [4]. More advanced driverless vehicles were developed by multiple research teams for the 2004 and 2005 DARPA Grand Challenges and the 2007 DARPA Urban Challenge (DUC), which accelerated the field's progress [5].

Some business groups have been pushing the boundaries of autonomous driving (AD) on urban roadways alongside academia, leading to significant development in recent years. When it comes to AVs, Google has the greatest experience. The corporation has logged over 2 million miles on its fleet and aims to utilize 100 of them in a trial project for the Ministry of Defense. Tesla is ahead of the curve in the auto industry because they included an autopilot option to their 2016 Model S cars. The mobility service provided by Uber has grown to compete with taxis in many cities across the world, and the business has lately revealed plans to replace its whole human-driven fleet with AVs [6]. Google's AV initiative, now called Waymo and a division of Alphabet, was covertly launched in 2009. Waymo presented a prototype of a fully AV in 2014. There are no controls of any kind in this automobile. Twenty million kilometers have been driven by Waymo vehicles on public roads across 25 cities in the United States [7]. After announcing a cooperation with the Shannon, Ireland-based autonomous car center, Jaguar Land Rover (JLR) Ireland has committed to using Ireland's 450 km of motorways to test its next-generation AV technology in 2020 [8].

2 Theoretical Definitions

2.1 Automation Levels and Architecture

For customers, the SAE established the J3016 "Levels of Driving Automation" specification. From no driving automation (level 0) to full automation (level 5) there are the six tiers of driving automation defined by the J3016 standard [9]. Figure 1 presents a high-level overview of the groups, widely employed in the auto industry for the safe creation, testing, and rollout of AVs [10]. Autopilot and Traffic Jam Pilot are only two examples of the SAE level 2 automation capabilities that car manufacturers like Audi and Tesla have implemented. Since 2016, Alphabet's Waymo has considered a business model based on SAE level 4 self-driving taxi services that may generate fares in a specific region of Arizona, USA [1].

Figure 2 depicts the relationships between the vehicle's fundamental competencies and its interactions with its surroundings, broken down into three categories: environmental perception, planning and decision, and motion and vehicle control. The basic hardware and software components and their implementations are described in Fig. 2A from a technical viewpoint, and Fig. 2B provides a functional analysis of the relationship between the four major functional units and the information flow.

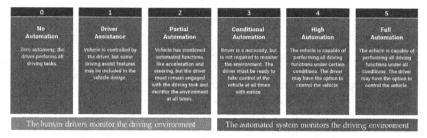

Fig. 1. The SAE Levels of Automobile Automation [10]

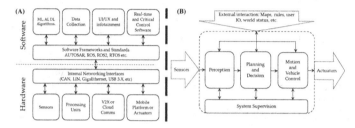

Fig. 2. Automated Driving System Architecture [11]

Perception is the capacity of an autonomous system to gather information and extract pertinent knowledge from its surroundings. In order for a robot to accomplish its higher-level goals, such as moving a vehicle from its current location to a designated destination while avoiding obstacles and maximizing its performance according to a set of established heuristics, the robot must engage in a process known as planning. Control competence, meanwhile, is the robot's ability to carry out the tasks that have been generated by more abstract processes [6].

2.2 Dataset

The acquisition of datasets is imperative for conducting adverse weather re-search. The extraction of features from datasets is a crucial step in object detection tasks, and the evaluation and validation of algorithms typically require the use of datasets. Sufficient data encompassing various weather conditions is imperative for effectively addressing the challenges posed by inclement weather in AD. Regrettably, a significant proportion of the datasets that are frequently utilized for training purposes lack a substantial number of distinct conditions beyond those associated with unobstructed atmospheric conditions. Table 1 enumerates a number of prominent datasets that have been utilized by scholars in the present study.

Table 1. Datasets for AD Systems in Adverse Condition [12, 13]

Year	Dataset	Weather Condition				Night	Sensors
		Snow	Rain	Fog	Glare		
2022	Boreas	✓	✓	✗	✗	✓	LiDAR, 360° Radar, Camera, GNSS-INS
2021	CADCD	✓	✗	✗	✗	✗	LiDAR, 8 Cameras, GNSS, and IMU
	WADS	✓	✓	✗	✗	✓	3 LiDARs, 1550 nm LiDAR, Camera, LWIR Camera, NIR Camera, GNSS, IMU
	GROUNDED	✓	✓	✗	✗	✗	LiDAR, Camera, RTK-GPS, and LGPR
	Radiate	✓	✓	✓	✗	✓	LiDAR, 360° Radar, Stereo Camera, GPS
	HSI-Drive	✗	✓	✓	✗	✓	Photon focus 25-Band Hyperspectral Camera
2020	LIBRE	✗	✓	✓	✓	✗	Camera, IMU, GNSS, Can, 360° 4K Cam., Event Cam., Infrared Cam.
	nuScenes	✗	✓	✗	✗	✓	LiDAR, 5 Radars, 6 Cameras, GNSS, IMU
	DAWN	✓	✓	✓	✗	✗	Camera (Internet-Searched Images)
	Waymo Open	✗	✓	✗	✗	✓	5 LiDARs + 5 Cameras
2019	ApolloScape	✗	✓	✗	✓	✓	Two LiDARs, Depth Images, GPS/IMU
	EuroCity	✓	✓	✓	✗	✓	Two Cameras
	D2-City	✓	✓	✓	✗	✗	Dashcams

3   Related Works

Hussain et al. [14] separated the implementation and design problems into distinct groups, such as price, software complexity, digital map constructing, simulation, and validations. Issues of security, privacy, and the management of scarce resources were discussed. But they investigated AD only from a social or non-technical perspective. Using six requirements as parameters, Yaqoob et al. [15] emphasized the progress made in AD research and assessed the future obstacles in the field. Failure tolerance, low latency, sound architecture, careful management of resources, and safety are fundamental needs. Challenges to deep learning-based vehicle control were outlined, and an overview of deep learning approaches and methods for controlling AVs was provided by Kuutti et al. [16]. In-depth discussions were held about the best ways to implement these strategies for controlling the vehicle laterally (steering), longitudinally (acceleration and braking), and simultaneously (lateral and longitudinal control). In their survey of the field of AD architecture, Velasco-Hernandez et al. [11] distinguished between technical and functional designs. The perceptual stage was then highlighted as an important aspect of the solutions for AD, with the sensing component and sensor fusion techniques for localization, mapping, and obstacle detection being explained. In their article, Wang et al. [17] discussed advances in sensing technology and how they function under different circumstances. In addition, they offered methods for establishing motion models and data associations for multi-target tracking and analyzed and summarized the current state of multi-sensor fusion solutions. To better grasp the effects, policy concerns, and planning issues, Faisal et al. [18] surveyed the current body of knowledge and identified potential areas of research neglect. It argues that urban areas must be ready for the advent of AVs; however, the choice of keywords for a search could exclude some relevant publications; analytical techniques could have been used, but instead a manual literature review is conducted.

4 Research Analysis

4.1 The Sensor Ecosystem for AVs

Light Detection and Ranging (LiDAR)
LiDAR was created in the 1960s and has since been used extensively in aerial and space-based topographical mapping. About halfway through the 90s, companies that specialized in making laser scanners began shipping the first commercial LiDARs, which had a pulse rate of between 2,000 and 25,000 PPS and were used for topographical mapping [19]. One of the core perception technologies for ADAS and AD cars is LiDAR, the evolution of which has been constant and substantial over the past few decades. LiDAR, a form of remote sensing, functions by sending out pulses of infrared or laser light and picking up any reflected energy from the targets of interest. The device picks up on these echoes, and the time it takes for the light pulse to travel from emitter to receiver provides a distance measurement. LiDAR sensors are in greater demand to support the developing field of autonomous robots, drones, humanoid robots, and AVs due to their performance features, which include measurement range and precision, robustness to environmental changes, and quick scanning speed. Many new LiDAR sensor businesses have cropped up in recent years, and they have been busy releasing cutting-edge products in response to the industry's growing needs. The automotive LiDAR industry is expected to generate $6,910,000,000 by 2025. Equations 1 and 2 show how to use this electronically measured round-trip ToF (Δt) to determine how far away the reflection point is [20]:

$$P(R) = P_0 \rho A_0 \left(\pi R^2\right)^{-1} \mu_0 exp - 2\gamma R \tag{1}$$

$$R = (2n)^{-1} c \Delta t \tag{2}$$

In these equations, P_0 represents the optical peak power of the emitted laser pulse, ρ represents the reflectivity of the target, A_0 represents the aperture area of the receiver, μ_0 represents the spectral transmission of the detection optics, γ represents the atmospheric extinction coefficient, c represents the speed of light in a vacuum, and n represents the index of refraction of the propagation medium (~1 for air). There are two main types of LiDAR sensors, mechanical and solid-state (SSL). In the realm of AV R&D, the mechanical LiDAR is the most widely used long-range environment scanning option. To focus the lasers and record the desired field of view (FoV) surrounding the AV, it employs high-quality optics and rotary lenses powered by an electric motor. The horizontal FoV can be rotated to encompass the whole area around the vehicle. In contrast, SSLs do not rely on mechanical parts like moving lenses. Several micro-structured waveguides are used by SSLs to steer the laser beams so that they may gather environmental information. The resilience, stability, and generally lower costs of these LiDARs make them a viable option to their mechanical counterparts, the spinning LiDARs, in recent years.

Nevertheless, unlike conventional mechanical LiDARs, their horizontal field of view is often 120 degrees or less [21]. The features of a typical LiDAR sensor are listed in Table 2 [1, 22].

According to Table 1, modern LiDAR sensors used in AVs often have a wavelength of 905, making them the safest sorts of lasers (Class 1), with reduced absorption water compared to, say, 1550 nm wavelength sensors [20]. 1D, 2D, and 3D LiDAR sensors are the three main types of this versatile technology. One-dimensional (1D) sensors collect data about the world around them exclusively in terms of distance (in x coordinates). Further information regarding the target object's angle (y-coordinates) can be gleaned using 2D, or two-dimensional, sensors. In order to determine an object's height (in z-coordinates), 3D (or three-dimensional) sensors project laser beams horizontally and vertically. Point cloud data (PCD) in 1D, 2D, or 3D space along with intensity information of objects is the data type most commonly produced by LiDAR sensors.

Table 2. Comparison of LiDAR Sensors

Principle	Model	Channels	FPS (Hz)	Wavelength (nm)	Range (m)	VR(°)	HR(°)	VFOV(°)	HFOV(°)
Mechanical	Hesai XT-32	32	10	905	120	1	0.18	31	360
	Ouster OS1–64	64	10,20	850	120	0.53	0.7,0.35,0.17	33.2	360
	Robosense RS-32	32	5,10,20	905	200	0.33	0.1–0.4	40	360
	LeiShen C16-700B	16	5,10,20	905	150	2	0.18,0.36	30	360
	Velodyne Alpha Prime	128	5–20	903	245	0.11	0.1–0.4	40	360
	Hokuyo YVT-35LXF0	-	20	905	35	-	-	40	210
Solid State	Cepton Vista X90	-	40	905	200	0.13	0.13	25	90
	IBEO LUX 8L	8	25	905	30	0.8	0.25	6.4	110

The PCD for 3D LiDAR sensors includes data on the location and strength of barriers in the scene or environment. High-resolution laser pictures (or PCD) are typically generated by using 64- or 128-channel LiDAR sensors for AD applications [23]. Compared to the 32-, 64-channel LiDAR, the 128-channel LiDAR clearly provides more detail in the point clouds shown in Fig. 3.

Because to their superior field of vision, detection range, and depth perception, 3D spinning LiDARs are increasingly being used in autonomous cars. As the AVs gather information, a point cloud representing a dense 3D space (also called "laser image") is generated. Several sensors feed into the PCD, making sensor fusion a common practice.

Fig. 3. LiDAR Point Clouds from 32-, 64- and 128-Channel OS1 LiDAR [24]

This is due to the fact that LiDAR sensors, unlike camera systems, provide no color data about the environment.

Camera

Cameras are widely used as a means of perceiving one's environment. A camera captures an image of its surroundings by sensing light striking a photosensitive surface (the "image plane") through a lens (positioned in front of the sensor) [15]. Low-cost cameras with the right software may produce high-resolution photos of the environment and detect both moving and stationary impediments within their range of vision. In the case of road traffic vehicles, these features allow the perception system to recognize road signs, traffic lights, road lane markings, and barriers; in the case of off-road vehicles, a wide variety of objects can be detected. Monocular cameras, binocular cameras, or a combination of the two may be used in an AV's camera system. Monocular camera systems only use a single camera to capture multiple images. Traditional RGB monocular cameras are inherently limited in comparison to stereo cameras due to a lack of native depth information. However, depth information may be generated in specific applications or with more modern monocular cameras employing the dual-pixel autofocus hardware; hence, AVs typically employ a binocular camera system, which consists of two cameras mounted side by side. The stereo camera, often known as a binocular camera, uses the "disparity" between the slightly different images formed in each eye to simulate the way in which animals perceive depth. The two image sensors in a stereo camera are placed at different distances from one another. Different stereo cameras have different values for the distance between their two image sensors, which is referred to as the "baseline" [25]. Fisheye cameras are another type of camera used in AVs for environmental perception. Using only four cameras, a fisheye camera can provide a full 360-degree view, making it ideal for near-field sensing applications like parking and traffic congestion help [26, 27].

Imperfections in a camera lens' geometry can cause pincushion distortion, barrel distortion, and moustache distortion, all of which fall under the umbrella term "optical distortion." Such distortions might cause miscalculations in the predicted positions of detected barriers or features in the image. Thus, "intrinsically calibrating" the camera to estimate its parameters and correct its geometric errors is often necessary. Another possible drawback of cameras is the need for a lot of processing resources during image analysis. In light of the foregoing, cameras are a pervasive device that captures detailed films and photographs, complete with information about the observed environment's colors and textures. Camera data is frequently used in AVs for tasks such as traffic sign recognition, traffic light recognition, and road lane marking detection. As camera performance and the creation of high-fidelity images are so heavily reliant on ambient conditions and illumination, image data are routinely combined with other sensor data, including as radar and LiDAR data, to give dependable and accurate environment perception in AD.

MMW Radar

In order to determine the distance to an object, radar, which was initially developed before World War II, sends out a pulse of electromagnetic (EM) radiation into the region of interest and picks up the reflected signal, which can then be processed further. It does this by employing the Doppler effect of electromagnetic waves to calculate the velocities and locations of identified impediments [21]. The Doppler effect, also called the Doppler shift, refers to frequency variations or shifts caused by the relative mobility of a wave's source and its targets. For instance, as the target approaches the radar system, the frequency of the received signal rises (shorter waves). Doppler shift in radar frequency can be described mathematically as [28]:

$$f_D = 2V_r f C^{-1} = 2V_r \lambda^{-1} \tag{3}$$

In this equation, f_D represents the Doppler frequency in Hertz, V_r represents the relative speed of the target, f represents the frequency of the transmitted signal, C represents the speed of light (3×10^8 m/s), and λ represents the wavelength of the emitted energy. Doppler frequency shifts occur in a radar system twice: first when electromagnetic waves are sent out to the target, and again when the reflected waves return to the radar system (source). Current commercially available radars use frequencies of 24, 60, 77, and 79 GHz (millimeter wave or MMW). The 24 GHz radar sensors are expected to be phased out in the future due to their worse range, velocity, and angle resolution compared to the 79 GHz radar sensors, which causes difficulties in identifying and reacting to various risks [10]. Given that electromagnetic wave transmission (radar) is unaffected by atmospheric circumstances like fog, snow, or clouds, these systems can operate reliably regardless of the time of day or weather. One of the problems with radar sensors is that they have trouble differentiating between moving and stationary objects, such as road signs and guardrails, which can lead to false alarms. Radars may have trouble telling the difference between, say, a dead animal and the road because their Doppler shifts are so close [17]. Using the 79 GHz vehicle radar sensor (SmartMicro) in the research region yielded a significant percentage of false positive detections within the area of interest, according to preliminary results. Objects as close as 7–10 m from the installed sensors might cause false positive detections, as seen in Fig. 4. In the point cloud visualization, LiDAR point cloud data is represented by colored points, whereas radar point cloud data is represented by white points. At the range of 5 to 7 m from the radar sensor, the grey rectangle represents multiple false positive radar detections. Because the radar sensor is currently set to its short-range mode (maximum detection range of 19 m), it is unable to detect the traffic cone 20 m away.

Vehicle-mounted radars can take on a variety of waveforms, the most common of which being frequency-modulated continuous-wave (FMCW) and pulse radars. While a high-power signal will be transmitted in a brief, continuous cycle, pulse radar necessitates rigorous isolation of the transmitted signal in order to receive the echo signal,

Fig. 4. Visualization of False-Positive Detections [1]

leading to a high need for hardware and a complex structure. As a result, most mobile millimeter-wave radar systems use FMCW for transmission. FMCW radar allows for the simultaneous availability of target distance and relative speed, as well as the tunable resolution of both speed and distance [17]. Resolutions R_{res} and V_{res}, in terms of distance and speed, are as follows:

$$R_{res} = c(2B)^{-1} \tag{4}$$

$$V_{res} = \lambda\left(2T_f\right)^{-1} \tag{5}$$

assuming c as the light speed, B as the chirp bandwidth, λ as the wavelength, and T_f as the pulse duration. The target angle can be detected by using multiple receiving antennas, with the resolution of the angle depending on the actual angle between the target and the direction of radiation. Following is a formula for determining the angular resolution θ_{res}:

$$\theta_{res} = \lambda(NdCos(\theta))^{-1} \tag{6}$$

where N represents the total number of antennas, d represents the distance between any two antennas, and θ represents the true target azimuth. However, because of limitations imposed by the radar gear (such as transmit power and sample rate), the greater the maximum effective working distance, the less precise the parameters can be.

Common places for radar sensors in AD vehicles include the roof, under the bumpers or brand logos, and the top of the windshield. Any angular misalignment in the mounting positions and orientations of radars during manufacture is a major cause for concern, as it could lead to the radars missing or misreading obstructions in the vehicle's path, which could have catastrophic effects [29]. Vehicle radar systems can be broken down into three distinct groups: medium-range radar (MRR), long-range radar (LRR), and short-range radar (SRR). SRR is used for adaptive cruise control and early detection, MRR for side/rear collision avoidance system and blind-spot detection, and LRR for packing assistance and collision proximity warning in AV production. Radar sensors are widely employed in AVs to provide dependable and accurate perception of obstacles at all times of day and night, and they are among the most popular in autonomous systems because of their operability in low-light and bad-weather environments. Depending on the settings, it can conduct short-, medium-, or long-range mapping. It also offers extra data like the observed speed of moving obstructions. However, due to their lower resolution than cameras, radar sensors are rarely used for object detection. To make up for radar's shortcomings, AV researchers frequently combine radar data with that from other sensors like cameras and LiDAR.

Global Navigation Satellite System

Robots, AVs, unmanned aerial vehicles, aircraft, ships, and even smartphones all need navigation or positioning systems as a fundamental component of their technology. The Global Navigation Satellite System (GNSS) is a worldwide network of satellite navigation and positioning systems that includes the United States' Global Positioning System (GPS), Russia's GLONASS, China's BeiDou, Europe's Galileo, and others. The L-Band (1 to 2 GHz) that GNSS uses has minimal path attenuation while traveling through atmospheric obstacles like clouds and rain. One or more antennas, reprogrammable GNSS receivers, CPUs, and memories are all part of a GNSS sensor. Combinations of GNSS and RTK (real-time kinematic positioning) systems that relay correction data via ground base stations are common [29]. The most well-known GNSS system is the United States' Global Positioning System (GPS), which offers positioning, navigation, and timing (PNT) services. Due to its accessibility, accuracy, and ease of use, GPS has become an integral part of the global information infrastructure. In the early 70s, the United States Department of Defense (DoD) began developing GPS. The system has three main components: the space segment, the control segment, and the user segment. The United States Air Force creates, maintains, and runs the GPS system's space and control parts [30].

The receiver's ability to detect at least four satellites, calculate the distance to each, and then use trilateration to estimate its own location is central to the GNSS's functioning principle. The accuracy of GNSS can be compromised by a number of factors, including timing problems due to discrepancies between the satellite's atomic clock and the receiver's quartz clock, signal delays due to ionospheric and tropospheric propagation, the multipath effect, and satellite orbit uncertainties [10]. Current vehicle positioning systems make use of a combination of satellite data and data from a wide variety of on-board sensors, including inertial measurement units (IMUs), LiDARs, radars, and cameras, to determine precise location.

Ultrasonic Sensors

Several industrial detecting activities can be accomplished with the help of ultrasonic sensors. They can detect anything from solids to liquids to granules to powders. For use in automobiles, ultrasonic sensors require sonic transducers to transmit sonic waves. The ultrasonic sensor is one type of popular automobile sensor that is rarely discussed in ADS modalities. Ultrasonic has been the most reliable and cost-effective sensor for parking assistance and blind spot monitoring, and it is widely used on bumpers and the entire automobile body. Ultrasonic sensors work on a similar basis to radar in that both use the time it takes for an emitted electromagnetic wave to travel to calculate a distance, but ultrasound operates in the much higher frequency range of 40 kHz to 70 kHz. As a result, the effective range of ultrasonic sensors is typically no more than 11m, limiting its utility to close-range applications such as backup parking [31]. In fact, ultrasonic technology is used in Tesla's "summon" feature to guide the vehicle through parking lots and garage doors [32]. Transmission-to-reception time-of-flight (ToF) measurements are the backbone of most modern ultrasonic sensors. Distance (d) to an object or reflector within the measuring range is then determined using the measured ToF, as illustrated in Eq. 7.

$$d = Speed\ of\ Sonic\ Wave \times ToF \times 2^{-1} \qquad (7)$$

Sound waves have a speed of 340 m per second in air; however, this value varies with factors including atmospheric pressure and humidity (the speed of sound rises by 0.6 m per second for every degree Celsius). Sonic waves take about 3×10^{-3} s to travel 1 m, compared to 3.3×10^{-9} s for light and radio waves. Because of these significant speed variations, ultrasonic systems can make use of low-speed signal processing. Yet, due to the impact of atmospheric pressure on ultrasonic sensors' overall performance, SRRs and other technologies are often used.

4.2 Autonomous Driving

Ego-Position
When first introduced, intelligent transportation systems included AV technology as a central tenet of their framework. Major advancements have been achieved in this area in recent years, and several businesses have begun conducting field tests. Knowing one's own location (ego-position, or self-localization) to within a few centimeters is essential for AD [33]. Even while GNSS can achieve this degree of precision in clear skies, the positioning quality suffers greatly in congested metropolitan environments due to signal blockage and the multipath effect. Vision is one of the technologies that can supplement or replace GPS. Despite significant advancements, vision-based approaches are still hindered by things like weather, variations in lighting, and shadows. LiDAR has lately gained traction as the major technology for the perception of AVs; this is because numerous businesses are attempting to reduce its costs. LiDAR provides precise readings because of its high resolution, great range, and insensitivity to ambient light. Numerous automakers have already begun installing this technology in their commercial vehicles for use in a variety of driver aid functions. Localization using LiDAR can be broken down further into two categories: SLAM (simultaneous localization and mapping) and map-based methods. There are two primary types of SLAM techniques, namely feature-based and scan-based. The accumulated errors in SLAM-based localization are a common issue. The solution to this issue is to often update the location with the help of global features [34]. Inside or outdoor, SLAM is a broad and significant topic in today's robotic and smart industry. Smart autonomous mobile robots rely on navigation, localization, and mapping technology. Safe or high-risk and difficult-to-navigate, SLAM plays a significant part in smart manufacturing. Leonard and Durrant-Whyte [35] conducted pioneering work in SLAM in the early 1990s, and the field has since expanded at a rapid pace. Figure 5 portrays the SLAM captured through the utilization of REV7 OS2 technology.

Fig. 5. Palm Drive, Stanford University, SLAM Data Acquired with REV7 OS2 [36]

Map-Matching Technology

The complexity of AD is not diminished by the provision of an HD map alone. To make use of a map, a precise estimate of the vehicle's location is required. GNSS was commonly used to find one's own location in the past. Nevertheless, when applied to automated driving, GNSS's position-measuring abilities cause serious problems. GNSS technologies fall short of the precise location accuracy needed for AVs in a few different conditions. Both forested and open spaces, such as rural areas, fall under this category. When towering buildings are closely spaced in a metropolis, for instance, the interference caused by multiple GNSS signal channels renders precise position measurement challenging. Also, location information is unavailable inside a tunnel due to a lack of satellite signals. Coupling GNSS and INS is commonly used to lessen the impact of such issues. Accuracy, however, suffers in areas where GNSS signals are intermittently unavailable. The other issue is that if the digital map itself contains a substantial mistake, then even if real-time kinematic GNSS (RTK-GNSS) can estimate the vehicle's position to the centimeter level, the vehicle and map cannot be brought into perfect correspondence with one another. Thus, it is crucial to make a real-time estimation of the vehicle's location on the map. The vehicle's location is estimated using map-matching technologies [37].

For the purpose of self-localization, many studies have employed the usage of a map-matching technique, comparing the user's position to that of a reference digital map and sensor data. There are precise coordinates for the sensor features around roads on the reference map. Predefined maps often make use of one of three map types: 3D point cloud map, 2D image map, or vector structured map. A 3D point cloud map details the 3D topography and features of the area around a given road. This style of map has a low ongoing cost but a massive data footprint. The road's surface might be reconstructed from a 3D point cloud and then used to generate a 2D picture map. This results in a significantly smaller data set when compared to a traditional 3D map. In a vector map, polynomial curves act as road and lane boundaries in place of the traditional white lines and curbs. This map is small in terms of the amount of data it stores, but it requires a lot of upkeep. Recently, 3D Planar Surface Map, as shown in Fig. 6, has been proposed, which is superior to the previous three models since it delivers more information than a 2D image map while taking up less space in storage, rendering it ideal for densely populated places like metropolis.

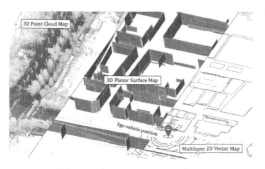

Fig. 6. Vehicle Self-Localization Using the 3D Planar Surface Map [33]

Due to the sensor's excellent measurement precision and resilience to day/night transitions, a LiDAR-based approach was investigated to achieve decimeter-level accuracy. A histogram filter and scan matching are two examples of the map matching techniques suggested in similar publications. These techniques often make an estimate of the vehicle's location by using the color of the road paint and the shapes of nearby buildings as landmarks. Several camera-based approaches that map observed images into a 3D LiDAR surface are proposed for inexpensive deployment [38].

Fig. 7. Localization Quality for Each LiDAR, per Route, Color-Coded to Show Intensity [39]

The qualitative outcomes of the various LiDARs' localization efforts are depicted in Fig. 7. After being modified, the original cloud from each LiDAR is depicted here, demonstrating the extent of coverage that was attained. However, alternatives to LiDAR-based self-localization are also being studied. The position is estimated using either the image features or the MMW radar characteristics; however, at the moment, the resulting location is less accurate than the LiDAR-based approach.

Environmental Perception

In real time, onboard sensors must detect and name road features like traffic signals, nearby obstacles, and moving traffic participants like cars, pedestrians, and cyclists. A self-driving car needs to be able to recognize road features along the lanes it's driving on so it can stay under the speed limit. Static road characteristics can be disregarded in a digital map, such as speed limits and stop lines, if the information is entered beforehand. However, identification of dynamic road components like traffic lights requires near-instantaneous processing. Thus, the traffic light is one of the most important constantly changing aspects of roads at crossings. Due to the need to classify the illumination hue, a color camera is required to identify the traffic lights. It is necessary to be able to recognize the condition of the traffic light from a distance of more than one hundred meters in order to drive safely through junctions. Incorporating a digital map and an earlier map-based detection method enables the determination of search zones to limit the amount of false positives. No matter what sensors are used, extracting the road surface and detecting objects on the road are two crucial parts of the perception task [6].

Role of Sensors in Perception

Accurate measurement precision is required for the detection of stationary obstacles in the vehicle's path. Range sensors like LiDAR, stereo camera, or microwave radiometer (MMW radar) are typically used to detect barriers in the environment. By employing binary bayes filter as time-series processing, a two-dimensional or three-dimensional static obstacle map with empty space can be constructed from which instantaneous erroneous detection can be mitigated to some extent. In addition, A machine learning classifier is fed the range sensor's object form and the camera's image data to determine the identities of the nearby humans and vehicles. The range sensor's observation points are clustered, and then the object category is classified from the features of each shape using ML approaches [11]. A dense observation point cloud can be acquired at a distance of a few tens of meters from the item, allowing for confirmation of its exact outline, but the sparseness of the resulting point cloud presents a challenge in that it is hard to extract fine shape information for faraway objects. However, when using camera images for recognition, more dense observation information may be collected in comparison to LiDAR, allowing for the recognition of objects further away than 100 m with the proper lens choice. Now, with the advent of GPU-based acceleration, a very accurate camera-based recognition approach called deep neural network (DNN)-based recognition has been developed [40]. Rectangular bounding boxes in the image are a common result of these methods of recognition, making it difficult to get distance information of the object directly. In order to determine where the detected objects are in space, one must use sensor fusion with a stereo camera or other range sensor to get distance information.

Predicting the future state of objects, in addition to recognizing people in the road, is crucial for collision detection, synchronized driving, and other driving tasks such as maintaining a safe spacing between vehicles; hence, not only the locations, but also the speeds and accelerations of the moving objects need to be estimated. Probabilistic methods like the Kalman filter and the particle filter are commonly used to estimate the states of nearby objects. Range sensors like LiDAR, MMW radar, and stereo cameras are typically employed due to the necessity of determining the distance to the target item [41]. Using a combination of sensors, rather than just one, can lead to more accurate object tracking. An identified object's future path can be effectively predicted using the digital map's information on road shapes. Not only can the road's structure and connection information be used to anticipate an object's behavior in a few seconds from now, but also its current motion status. In addition, implementing traffic rules and seeing how people in the area respond to one another can help one anticipate and prepare for more suitable behavior [42].

Trajectory Tracking
In order to make decisions during AD, three forms of situational judgment are needed. Route planning is the first technological advancement. This is standard fare for GPS systems in cars and other vehicles, but in this case, the journey must be plotted out lane by lane. In a setting where road information has been collected from a digital map, dynamic programming can be utilized to search the path along the connections of roads. If there is no road signage, as in a parking lot or a large area, the best path must be determined from the drivable region [43]. The second one is a situation of transportation that follows all applicable regulations. When going along the route, it is crucial to observe all applicable traffic laws. In order to provide a safe procedure when making an entrance judgment at an intersection, the recognition results of the traffic light status and the stop line position, in addition to the positional relationship of the oncoming vehicle and pedestrians on the zebra crossing, must be taken into account. Before approaching an intersection, drivers should also think about the relative importance of the current road and their intended destination; hence, it is crucial to make a good decision based on a rational analysis of the situation while driving in heavy traffic. Optimizing a trajectory is the third technological advancement. Trajectory planning is carried out to find the lowest-cost, collision-free path along the gathered route. The application of polynomial functions in trajectory planning allows for the generation of smooth trajectories with little acceleration (jerk) [37]. Public road driving behavior can be modeled as a decision-making system, allowing for highly adaptable AD.

Overview of the Self-Driving Vehicle Process
The procedure of processing sensor data is depicted in Fig. 8. AD requires the on-board computer to process a large amount of data. According to Fig. 6, AVs utilize LiDAR for a variety of purposes, including geolocation, traffic participant detection, and object estimation. Self-localization and the estimate of static and moving objects are two further applications of MMW radar in AV. Detecting traffic lights, identifying people in traffic, and estimating the distance to static and moving objects are just some of the many uses for cameras in AVs.

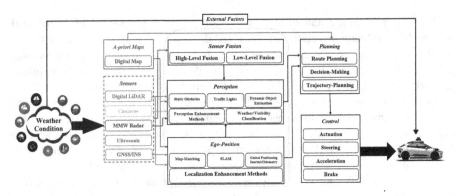

Fig. 8. Self-Driving Vehicle Process

4.3 AV Performance Under Challenging Circumstances

Adverse Weather

Because of the snow, the LiDAR sensor's peripheral vision is altered, the road surface is obscured, and there is more noise, and the MMW radar picks up some background noise, the visibility drops, and the road surfaces are obscured in the cameras [44]. GNSS systems are typically immune to atmospheric interference. These satellite networks operate at a frequency of roughly 1.575 GHz, which is fairly resistant to noise from the surrounding environment. Yet, Sometimes a GNSS receiver is unable to decode a complete navigation data string from satellites due to interference from windshield wipers. Inaccurate readings could be produced if the GNSS receiver failed to correctly decode the input string [45].To further the usefulness and operational scope of automated driving, addressing the issue of using it during winter is crucial. The vehicle's inability to self-localize is exacerbated by the fact that snow prevents accurate map data matching. Data collected during a snowstorm may not correspond to what is depicted on the map due to changes in the geometry of objects near the road and the invisibility of road paint. When this occurs, it becomes more difficult for an AV to choose the appropriate lane. There are three methods for maintaining accurate self-localization in the snow. In the first method, highly accurate RTK-GNSS is used. Pinpointing the position on the map can be difficult if the absolute accuracies acquired in real time and using the map are different. The second strategy is a technique for reconstructing observation data in the same way as rainfall measurements are reconstructed. It has been hypothesized that if the amount of snowfall is low, and the road surface and circumstances of the surrounding area can be viewed, even to a limited degree, then a more favorable outcome can be anticipated. The third strategy involves the utilization of a reliable snowfall sensor. MMW radars have been proposed for use in a system that uses self-localization to make the radars reliable sensors in bad weather. It has been asserted that the lateral precision of 0.2 m that can be reached with MMW radars is unaffected by snowfall; nevertheless, this is much less accurate than the lateral accuracy of 0.1 m that can be obtained with LiDAR when the weather circumstances are normal. The recognition of objects in snow is the next technical challenge for autonomous cars. Snowflakes in the air can just look like white noise to a LiDAR sensor as shown in Fig. 9 [37]. The perception of snow in the atmosphere is more pronounced compared to rain due to its relatively slower descent velocity and larger particle size. Various techniques have been devised to mitigate the problem of noise. Noise has been removed by reconstructing observational information in order to address this problem. By using machine learning to filter out irrelevant information from the sensor's observation, accurate recognition is possible [46]. Path planning on snowy roads is a separate technical challenge. Due to snowfall on the shoulder, the passable space on either side of the road may shrink or expand. When utilizing an HD map for AD, it can be challenging to make safe turns in the snow due to the fact that the center-lines are recorded with the assumption there is not going to be snow. Rather than relying purely on map data, it is recommended to use an adaptive plan for the drivable region that accounts for the present weather and road conditions while employing automated driving in snow.

Fig. 9. LiDAR Measurements in Wintry Weather, Rain Splash from Neighboring Car [37]

The LiDAR sensor's reflectivity decreases as rain falls, the recognition distance drops, and the splash causes the shape to shift. It also diminishes camera sensor visibility and shortens the range at which MMW radar can measure. In 2 mm/h rain, the 2000 m visibility that may be achieved with LiDAR on a clear day is reduced to about 1200 m for the 905 nm band and 900 m for the 1550 nm band. As the rain rate reaches 25 mm/h, visibility at 2000m decreases to 700m for wavelengths of 905 nm and 450 m for wavelengths of 1550 nm [43]. For 77 GHz MMW radar systems used in autonomous vehicles (wavelength $\sim$ 3.9 mm), Extreme rainfall of 150 mm/h reduces the detectable range to 45% from its normal value [44], but the effect of attenuation is not significant at short distances (from 0.0016 dB/m for 1 mm/h to 0.032 dB/m at 100 mm/h) [47]. The maximum detectable range, however, can be reduced by rain backscattering or rain clutter. Furthermore, rain clutter can surpass the detection threshold, leading to false alarms, because the received backscatter from the rain depends on R^2, rather than R^4 for the target echo. In addition to rain's impacts, puddle splash from approaching and nearby cars should be taken into account as shown in Fig. 9. It is difficult to tell if a

LiDAR observation is of a spray of water or an actual impediment that needs to be avoided. Sensor fusion, such as the incorporation of camera image data, is required in such a scenario. MMW radar is more resistant to the environment than LiDAR, yet it still suffers from radio attenuation when it rains. Although rain does not weaken reflections, it can generate artificial barriers, particularly during periods of non-uniform and heavy precipitation, as illustrated in Figs. 14A, B, and C. The majority of LiDARs identified the water droplets dispersed by the sprinklers as vertical pillars. Due to the instability of infrared reflectance properties during rain, its effect on self-localization performance must be taken into account. Inaccurate estimates may be made during map matching if the reflectance received from the map and the actual observation has different properties. Many techniques have been proposed as possible answers to these sorts of problems, including reconstruction of data and strong matching against the influence of noise. It is anticipated that improved localization will result from a narrowing of the gap between the existing map and the observation data [48]. Nonetheless, newer digital LiDAR sensors have been developed with improved performance in rainy conditions. Rainy (right) and dry (left) conditions captured by the OS1 LiDAR sensor are shown in Fig. 10. The OS1 LiDAR sensor created the three photos in the upper right corner as structured data panoramas without the aid of a camera. The upper image is the result of the LiDAR sensor's collection of natural light. Middle image shows the relative strength of the signal (strength of laser light reflected back to the sensor). The range between the sensor and the object is depicted in the third image. Water droplets on the lidar sensor glass appear not to affect the signal strength or range of the images. With less sunshine coming through the clouds, the ambient image is more grainy than usual, but the rain has had no discernible effect. The LiDAR sensor's range is diminished on the road's surface, which has unintended consequences. Its laser light is partially reflected by the road's wet surface and lost. This reduces the sensor's long-distance visibility of the road. Nonetheless, the sensor's range is unaffected by anything else in the room (cars, buildings, trees, etc.). The figure below, which shows the difference between a rainy and dry day, exemplifies this phenomenon. In the rain, visibility of buildings is unaffected, but vision of the road surface is diminished due to specular reflection [49]. Also, the IP68 and IP69K requirements have been met by OS1.

Fig. 10. Performance of OS1 LiDAR Sensor in Dry (Left) and Rainy (Right) Condition [50]

The updated L2X chip, as implemented in Rev 06, enhances the capacity of LiDAR sensor to detect objects in the presence of obscurants, including rain, fog, dust, snow, and even delicate structures such as wired fences. The utilization of both the primary and secondary returns of incoming light by sensors can enhance the precision of object detection in scenarios where objects are partially concealed by external elements such as foliage or smoke, delicate objects, and environmental obscurants like rain or fog. The limitations of cameras stem from their dependence on ambient light for pixel detection, rendering them incapable of detecting objects that are even partially obstructed. Ouster sensors differ from other sensors in that they employ a sizeable optical aperture and sophisticated digital processing through the L2X chip. This allows the LiDAR sensor to effectively navigate through obstructions such as rain, fog, and snow, which would otherwise impede the functionality of cameras. Neither the person in Fig. 11A nor the car in Fig. 11B can be seen by the GoPro camera due to the fog and dust, respectively, but both are captured by the Ouster OS1 Rev 06 sensor.

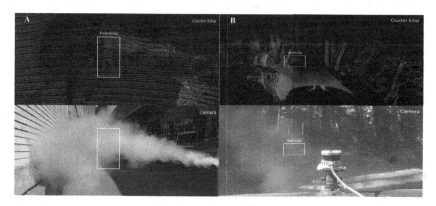

Fig. 11. A) Pedestrian in the Fog, B) Automobile on a Dusty Dirt Road [51]

Attenuation attributed to sun glint is hotly debated among sensors. The most affected technologies are LiDAR and cameras, while MMW radar, ultrasonic sensors, and GNSS are unaffected. As can be seen in Fig. 12, both the LiDAR data (point cloud colored by intensity) and the RGB camera image (top right inset) are badly impacted by bright light. In the experiment, the vehicle is positioned 40 m from the Xenon light source generating 200 klx. Classification is impossible due to weak object detection, four 3D points for the mannequin, cyan box, three points for the reflective targets, magenta box, and ten points for the black car, green box, but he thermal camera is not affected by the light and provides sharp images (top left inset).

Whiteouts on camera sensors are another effect of sun glare. Camera-only recognition functions, like traffic signal recognition, need to account for the likelihood of degraded performance. As a result, enhancements to both the algorithm and the hardware should be taken into account. High dynamic range (HDR) cameras, which can withstand sudden shifts in lighting without losing detail, have been created recently [37]. In particular, solving the challenge for intense sunlight from a software and hardware perspective is crucial since traffic light identification is a vital duty when determining the entrances of

Fig. 12. The Detrimental Impact of Intense Light on the Camera and LiDAR Sensor [29]

junctions. Cameras, LiDAR, and MMW radar can all be used to recognize pedestrians and other road users in the immediate vicinity. As was previously indicated, image recognition under sun glare is thought to suffer from increased instances of erroneous detection and non-detection. As a result, it is crucial to construct a system with numerous systems instead of relying on a single sensor [41].

Degradation to reflectance and shorter measured distances from the LiDAR sensor are two effects of fog. It also reduces the range at which MMW radar can measure and reduces the quality of images captured by cameras. While having the same maximum range performance in typical conditions, the 905 nm and 1550 nm wavelengths' visibility has decreased to around 200 m and 120 m, respectively. The camera's performance was evaluated in a fog chamber using particles measuring 2 microns and 6 microns in size. Visibility was only 30 m instead of the ideal 60 m [10]. One option for maintaining visibility in such conditions is to use an infrared camera. So, LiDAR and cameras are both impacted by fog; however, it has been observed that a deep neural network employing this sensor data might mitigate the performance decline. At shorter ranges, around 20 m, MMW radar was reported to have little to no impact on its ranging accuracy. Based on the data provided, it is clear that water vapor's low radio-signal-absorption coefficient has no effect on the detectability of millimeter waves [29]. Several analog LiDARs' qualitative data during severe weather are depicted in Fig. 14. Experiments with fog (top row), rain (30 mm/h) and strong light (bottom row) have been conducted with the vehicle positioned in close proximity to the targets and the rainy and bright areas, respectively. The following illustration serves as a visual summary of all the preceding discussion in this paper.

4.4 Sensor Fusion

The majority of autonomous systems, such as self-driving cars and unmanned ground vehicles, rely heavily on sensor fusion as an essential part of their operation. To reduce the number of detection errors and get over the constraints of individually operating sensors, it combines the data acquired from numerous sensing modalities. Moreover, sensor fusion allows for the creation of a consistent model that can accurately perceive the world in a wide range of conditions [53]. Industrial cameras, LiDAR, MMW radar,

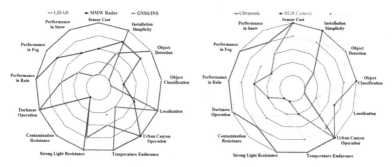

Fig. 13. Performance of the Sensors [10, 29]

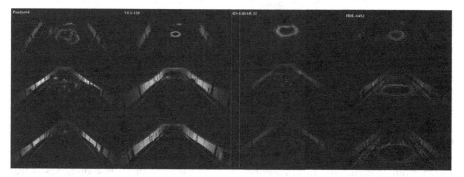

Fig. 14. Unfavorable Weather Detected by LiDARs, Wherein Hue Denotes Severity [52]

GPS, IMUs, etc. are only some of the sensors currently employed in smart vehicles for environmental sensing. These sensing systems not only gather the environmental data to be perceived, but also process the signals related to that data [54]. The environmental data collected by vehicle-mounted cameras, for instance, is often processed using artificial intelligence methods. New neural networks, such as DNN and F-RCNN, are regularly deployed to increase its processing speed and accuracy. In addition, visual data-based deep learning and semantic segmentation techniques are commonly applied in a variety of traffic contexts [55, 56]. In the Multi-source and heterogeneous information fusion (MSHIF) system, camera-radar (CR) is the most common sensor combination used for perception of the environment, followed by radar-camera-LiDAR (RCL) and camera-LiDAR (CL). High-resolution images are provided by the CR sensor set, and additional data on the distance and speed of nearby obstructions is gleaned. Tesla, for instance, used a variety of sensors, including ultrasonic ones, in conjunction with a CR sensor combination to get perception of the environment around the car's location. With the help of LiDAR point clouds and depth map data, the CLR sensor setup is able to deliver resolution at longer ranges and a thorough comprehension of its environment. Safety redundancy in the entire autonomous system is also enhanced [17]. Figure 15 depicts the typical locations, ranges, and uses of environmental sensors in AV systems.

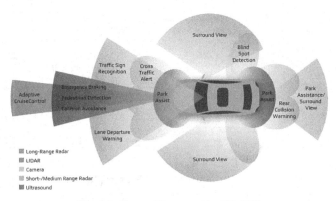

Fig. 15. Sensor Placement in AVs [57]

Methods of Sensor Fusion

High-level fusion (HLF) or semantic-level fusion is a process where multiple sensors independently run a detection or tracking algorithm and then combine the results to make a single, unified call. It is indeed simpler, uses less data transfer, and takes less processing power. In addition, HLF allows for a standardized interface towards the fusion method and does not require a deep understanding of the signal processing algorithms being used; however, it does not give sufficient data because classifications with a lower confidence value are ignored. There is also little to no influence on data precision or latency from changing the fusion algorithms. The adoption of HLF approaches is frequently attributed to their comparatively lower complexity compared to LLF and MLF approaches. The HLF system is deemed insufficient as it disregards classifications with a lower confidence value in cases where multiple obstacles overlap, thereby leading to a lack of comprehensive information. For higher quality and greater insight, sensor data can be incorporated in data-level fusion or low-level fusion (LLF) at the raw data level of abstraction. When multiple sensors work together, the data they collect is more precise (has a higher SNR) than it would be if each sensor worked alone. This suggests it may enhance detection precision. By bypassing the sensor's processing time, LLF reduces the time it takes for the domain controller to take action after receiving data. Particularly helpful in time-sensitive systems, this can boost performance. However, it produces copious amounts of data, which may strain storage capacity or network throughput. Also, it can be challenging to deal with poor readings when using LLF since it requires precise calibration of sensors to accurately blend their perspectives. Although it is possible to fuse data from several sources, the presence of redundant data significantly reduces the efficiency of this procedure. There are a number of difficulties that arise while putting the LLF strategy into effect. In mid-level fusion (MLF), or feature-level fusion, characteristics from many sensors are extracted based on their context and then fused to create a single signal for further analysis. It creates compact data sets with less computational overhead compared to LLF methods. To perform detection and classification, it combines features from multiple sensors into a single unified representation, such as color information from photos or position features of radar and LiDAR. In addition, the MLF method yields a robust feature vector, and algorithms for features selection can improve recognition accuracy by locating useful features and feature subsets, but locating the most relevant feature subset calls for substantial training sets. Extraction and fusion of data from sensors necessitates precise sensor calibration. Nevertheless, it appears that the MLF's weak understanding of the surroundings and loss of contextual knowledge prevent it from achieving an Level 4 or 5 AD system. [21, 58, 59].

State-of-the-art Perception Methods of Sensor Fusion

In order to forecast the geometry and semantic information of essential parts on a road, perception systems typically accept as input multi-modality data, such as images from cameras, point clouds from LiDARs, and HD maps. Results from high-quality perception are used to validate observations in subsequent processes including object tracking, trajectory prediction, and route planning. A perception system can encompass various vision tasks to achieve a thorough comprehension of driving environments. These tasks may include object detection and tracking, lane detection, as well as semantic and instance segmentation. Within the realm of perception tasks, the detection of three-dimensional objects is a crucial component of any AD perception system. The objective of 3D object detection is to anticipate the positions, dimensions, and categories of significant objects such as automobiles, pedestrians, and cyclists within a three-dimensional space. While 2D object detection solely produces 2D bounding boxes on images, disregarding the factual distance data of objects from the ego-vehicle, 3D object detection prioritizes the identification and positioning of objects within the tangible 3D coordinate system of the real world. The utilization of geometric information derived from 3D object detection in real-world coordinates can facilitate the measurement of distances between the ego-vehicle and crucial objects. Moreover, this information can aid in the development of driving routes and collision avoidance strategies. Sensors like radars, cameras, and LiDAR can all contribute raw data for 3D object detection. The objective of 3D object detection is to anticipate the bounding boxes of 3D objects in driving environments based on sensory inputs. The formula for detecting 3D objects in a given scene can be expressed as $\beta = f_{det}(I_{sensor})$, wherein $\beta = \{\beta_1, \ldots, \beta_N\}$ denotes a collection of N 3D objects, f_{det} represents a model for detecting 3D objects, and I_{sensor} refers to one or more sensory inputs. The representation of a 3D object B_i is a pivotal issue in this task, as it dictates the requisite 3D information for subsequent prediction and planning stages. Typically, a three-dimensional or 3D object is depicted as a 3D cuboid that encompasses said object. The cuboid is denoted as $B = [x_c, y_c, z_c, l, w, h, \theta, class]$, where (x_c, y_c, z_c) represents the 3D center coordinate of the cuboid. Also, l, w, and h denote the length, width, and height of the cuboid, respectively. The heading angle, or yaw angle, of the cuboid on the ground plane is represented by θ. Lastly, class parameter indicates the classification of the 3D object, such as cars, trucks, pedestrians, or cyclists.

For 3D object detection, raw data can be provided by radar, camera, and LiDAR. Radars have a wide detection arc and can function in a wide variety of climates. Radars, because of the Doppler effect, may also be able to measure relative velocities. Because of their low cost and widespread availability, cameras are increasingly being used to decipher semantics such as the type of traffic sign. For 3D object detection, cameras generate images $I_{cam} \in R^{W \times H \times 3}$, where W and H are the image's width and height and where each pixel contains three RGB channels. Although inexpensive, cameras have serious drawbacks when it comes to 3D object identification. To begin, cameras cannot directly get 3D structural information about a scene; they can only gather appearance information. However, 3D object detection typically necessitates precise localization in 3D space, and 3D information (such as depth) derived from photos is notoriously inaccurate. Furthermore, it was shown that image-based detection is particularly sensitive to environmental and temporal extremes. LiDAR sensors provide an alternate method for obtaining detailed 3D scene structures. Each range image pixel comprises the range r, azimuth a, and inclination $\emptyset$ in the spherical coordinate system in addition to the reflective intensity, thus a LiDAR sensor that emits m beams and performs measurements for n times in one scan cycle can generate a range image $I_{range} \in R^{m \times n \times 3}$. LiDAR sensors produce raw data in the form of range images, which, after being transformed from spherical to Cartesian coordinates, can be used to create point clouds. Each point in a point cloud has three channels of x, y, and z coordinates; therefore, $I_{point} \in R^{N \times 3}$ representation makes sense when N is the number of points in a scene. Point clouds and range pictures both contain precise 3D data collected by LiDAR sensors. Since LiDAR sensors are less susceptible to time and weather variations, they are preferable to cameras for 3D object detection. Popular perception system algorithms include Convolutional Neural Networks and Recurrent Neural Networks. YOLO, SSD, VoxelNet, Point Net, ResNet, and CenterNet are some of the most widely used foundation networks for sensor fusion techniques. Table 3 presents a thorough analysis of the performance of various

Table 3. Comparative Study of Three-Dimensional Object Detection Techniques

Fusion Level	Method	Inference Time	Year	nuScenes		KITTI		
				NDS	mAP	Easy	Med	Hard
Early	Fusion Painting	-	2021	70.7	66.5	-	-	-
	MVP	-	2021	70.5	66.4	-	-	-
	PointPainting	-	2020	-	-	82.11	72.7	67.08
	RoarNet	100	2019	-	-	83.95	75.79	67.88
	F-ConvNet	-	2019	-	-	85.88	76.51	68.08
	F-PointNet	-	2018	-	-	81.2	70.39	62.19
Late	Fast-CLOCs	125	2022	-	-	88.94	80.67	77.15
	CLOCs	150	2020	68.7	63.1	89.11	80.34	76.98
Middle	BEVFusion	-	2022	71.8	69.2	-	-	-
	UVTR	-	2022	71.1	67.1	-	-	-
	TransFusion	-	2022	71.7	68.9	-	-	-
	3D-CVF	75	2020	62.3	52.7	89.2	80.05	73.11
	EPNet	-	2020	-	-	89.81	79.28	74.59
	MVX-Net	-	2019	-	-	83.2	72.7	65.2
	MMF	80	2019	-	-	86.81	76.75	68.41
	ContFuse	60	2018	-	-	82.54	66.22	64.04

3D object detection techniques, encompassing mean average precision (mAP) and normalized detection score (NDS) on the nuScenes dataset, the inference time (measured in milliseconds), 3D vehicle detection on the KITTI test benchmark [13, 60–62].

Figure 16 depicts statistics on the characteristics, scenarios of each sensor, illustrating the efficacy of the MSHIF system in environment perception and target recognition.

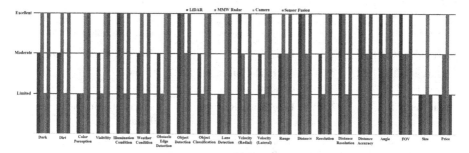

Fig. 16. Characteristics of Sensors and the Results of Fusion

Without a doubt, the extensive research behind multi-sensor fusion technologies has resulted in quite comprehensive benefits for autonomous systems, including everything from humanoid robots to AVs. These systems typically have a plethora of sensors that may provide copious amounts of data at a rapid rate. For instance, per hour, an AV might produce anything from 383.1 GB to 5.17 terabytes of data [63]; hence, a lot of computing resources are needed to analyze these data. Combining RL methods with supervised learning algorithms has the potential to lessen the software's need for processing speed, data storage, and time spent on training.

5 Discussion

The present study suggests a sensor fusion approach that integrates digital LiDAR, pair of stereo camera, thermal camera, 77 GHz MMW radar, and ultrasonic sensor. Table 4 displays other configurations of sensor fusion and their Objective.

As evidenced, LiDAR's efficacy is somewhat compromised in inclement weather conditions; however, digital LiDAR technology possesses the capability to operate in precipitation, such as rain and snow, as well as in dusty environments. Michaelis et al. employed defaced images generated from pre-existing datasets such as Pascal, Coco, and Cityscapes to assess the efficacy of state-of-the-art target detection algorithms [72]. Figure 17, presented below, illustrates a decrease in detection accuracy of no less than 31.1%, with a maximum decrease of 60.4% observed in certain scenarios. Thus, it is justifiable to infer that a solitary camera sensor would exhibit significant unreliability when subjected to harsh weather conditions. The efficacy of LiDAR and RGB camera is compromised in high luminance environments, whereas thermal camera is capable of accurate detection under such conditions. The utilization of MMW Radar has been observed to exhibit high dependability even in adverse weather conditions. Moreover, the issues associated with the Doppler effect have been effectively addressed through the

Table 4. Targeted Adversarial Conditions and Sensor Combination

Sensor Fusion	Year	Sensor Combination	Objective
MVDNet [64]	2021	LiDAR and Radar	Fog
SLS Fusion [65]	2021	LiDAR and Camera	Fog
Liu et al. [66]	2021	LiDAR and Camera	Fog, Rain, and Nighttime
John et al. [67]	2021	Thermal Camera and Visible Cameras	Low Light Conditions, and Headlight Glint
Rawashdeh et al. [68]	2021	RCL	Navigating in the Snow
Vachmanus et al. [69]	2021	Thermal Camera and RGB cameras	Snow
Radar Net [70]	2020	LiDAR and Radar	Potential Rain (NuScenes Dataset)
HeatNet [71]	2020	Thermal Camera and Two RGB cameras	Nighttime

integration of camera and LiDAR technologies. The ultrasonic sensor is often overlooked in the assessment of weather impacts; however, it exhibits distinct characteristics. The velocity of sound propagation in air is influenced by atmospheric pressure, moisture content, and thermal conditions. The variability in precision resulting from this phenomenon poses a significant challenge to AD, unless resorting to algorithms capable of adapting the measurements to the surrounding conditions, which entails additional expenses. However, ultrasonic technology possesses certain advantages, as its fundamental operation is less susceptible to adverse weather conditions in comparison to LiDAR and camera systems. The reliability of ultrasonic waves in low visibility environments surpasses that of cameras, particularly in areas with high glare or shaded regions beneath an overpass, as the return signal of ultrasonic waves remains unaffected by the target's dark color or low reflectivity. Also, the close-range capability of ultrasonic technology can be employed for the purpose of categorizing the state of the pavement. The identification of various types of roads such as asphalt, grass, gravel, or dirt can be achieved through the analysis of their back-scattered ultrasonic signals. It is reasonable to infer that this methodology can also be applied to distinguish snow, ice, or slurry on the road, thereby aiding in the classification of weather conditions for AVs. One additional benefit of using this sensor fusion is its capacity for ego-position, environmental awareness, and path planning in metropolitan regions characterized by towering edifices, congested thoroughfares, or subterranean passageways where GPS functionality may be compromised.

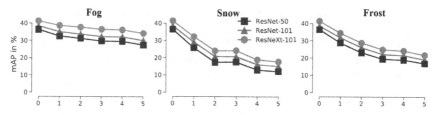

Fig. 17. The Impact of Various Forms of Corruption on Camera Accuracy [72]

6   Conclusion

Due to their individual strengths and weaknesses, none of the aforementioned sensors is sufficient on its own to meet the requirements of AVs, especially those at the level 3 and above; therefore, sensor fusion is emphasized as a means to take all circumstances into account. Naturally, it is crucial to implement LLF, or HLF, and to estimate the sensor's reliability. There are many obstacles to overcome in sensor fusion; problems with sensor fusion include noise resulting from missing values, calibration errors, loss of precisions, quantization errors, as well as overfitting of training datasets, biases in collected datasets, uncertainty in data measurements, and imprecision. Implementing sensor fusion may also be difficult due to the need to transform data from multiple sensors into a common reference frame. Also, the deployment of the automated driving system into society necessitates a system design based on a trade-off between the expense of installing the sensor in the infrastructure and the benefits of doing so. The adoption of AVs will help minimize accidents and traffic problems and make driving easier for everyone, and their progress and development are proceeding swiftly despite numerous obstacles, particularly the issue of cost.

References

1. Yeong, D.J., Velasco-Hernandez, G., Barry, J., Walsh, J.: Sensor and sensor fusion technology in autonomous vehicles: a review. Sensors **21**, 2140 (2021)
2. Autonomous Vehicle Market Size, Share, Trends, Report 2023–2032. https://www.precedenceresearch.com/autonomous-vehicle-market
3. 1918 March 10 Oakland Tribune. https://www.scribd.com/document/20618885/1918-March-10-Oakland-Tribune-Oakland-CA
4. Thorpe, C., Hebert, M.H., Kanade, T., Shafer, S.: Vision and navigation for the Carnegie-Mellon Navlab. IEEE Trans. Pattern Anal. Mach. Intell. (1988)
5. Urmson, C., Anhalt, J., Bagnell, D., et al.: Autonomous driving in urban environments: boss and the urban challenge. Springer Tracts in Advanced Robotics, pp. 1–59 (2009). https://doi.org/10.1007/978-3-642-03991-1_1
6. Pendleton, S., Andersen, H., Du, X., et al.: Perception, planning, control, and coordination for autonomous vehicles. Machines **5**, 6 (2017)
7. Glon, R., Edelstein, S.: History of self-driving cars milestones | Digital trends. https://www.digitaltrends.com/cars/history-of-self-driving-cars-milestones/
8. Jaguar Land Rover to partner with autonomous car hub in Shannon. https://www.irishtimes.com/business/transport-and-tourism/jaguar-land-rover-to-partner-with-autonomous-car-hub-in-shannon-1.4409884

9. Cui, G., Zhang, W., Xiao, Y., et al.: Cooperative perception technology of autonomous driving in the internet of vehicles environment: a review. Sensors **22**, 5535 (2022)

10. Vargas, J., Alsweiss, S., Toker, O., Razdan, R., Santos, J.: An overview of autonomous vehicles sensors and their vulnerability to weather conditions. Sensors (2021)

11. Velasco-Hernandez, G., Yeong, D.J., Barry, J., Walsh, J.: Autonomous driving architectures, perception and data fusion: a review. In: 2020 IEEE 16th International Conference on Intelligent Computer Communication and Processing (ICCP) (2020)

12. Huang, K., Botian, S., Li, X., et al.: Multi-modal sensor fusion for auto driving perception: a survey. arXiv:2202.02703 (2022)

13. Mao, J., Shi, Sh., Wang, X., Li, H.: 3D object detection for autonomous driving: a comprehensive survey. Int. J. Comput. Vis. (2022)

14. Hussain, R., Zeadally, S.: Autonomous cars: research results, issues, and future challenges. IEEE Commun. Surv. Tutorials **21**, 1275–1313 (2019)

15. Yaqoob, I., Khan, L.U., Kazmi, S.M.A., et al.: Autonomous driving cars in smart cities: recent advances, requirements, and challenges. IEEE Network **34**, 174–181 (2020)

16. Kuutti, S., Bowden, R., Jin, Y., et al.: A survey of deep learning applications to autonomous vehicle control. IEEE Trans. Intell. Transp. Syst. (2021)

17. Wang, Z., Wu, Y., Niu, Q.: Multi-sensor fusion in automated driving: a survey. IEEE Access **8**, 2847–2868 (2020)

18. Faisal, A., Yigitcanlar, T., Kamruzzaman, Md., Currie, G.: Understanding autonomous vehicles: a systematic literature review on capability, impact, planning and policy. J. Transp. Land Use 12 (2019)

19. The beginnings of LiDAR – A time travel back in history – Blickfeld. https://www.blickfeld.com/blog/the-beginnings-of-lidar/

20. Royo, S., Ballesta-Garcia, M.: An overview of Lidar imaging systems for autonomous vehicles. Appl. Sci. **9**, 4093 (2019)

21. Shahian Jahromi, B., Tulabandhula, T., Cetin, S.: Real-Time hybrid multi-sensor fusion framework for perception in autonomous vehicles. Sensors **19**, 4357 (2019)

22. Kim, J., Park, B., Kim, J.: Empirical analysis of autonomous vehicle's LiDAR detection performance degradation for actual road driving in rain and fog. Sensors. (2023)

23. Kodors, S.: Point distribution as true quality of LiDAR point cloud. Baltic J. Modern Comput. 5 (2017)

24. Li, L., Ismail, K.N., Shum, H.P.H., Breckon, T.P.: DurLAR: a high-fidelity 128-channel LiDAR dataset with panoramic ambient and reflectivity imagery for multi-modal autonomous driving applications. In: 2021 International Conference on 3D Vision (2021)

25. Garg, R., Wadhwa, N., Ansari, S., Barron, J.: Learning single camera depth estimation using dual-pixels. In: 2019 IEEE/CVF International Conference on Computer Vision (2019)

26. Yogamani, S., Hughes, C., Horgan, J., et al.: WoodScape: a multi-task, multi-camera fisheye dataset for autonomous driving. In: 2019 IEEE/CVF International Conference on Computer Vision (ICCV) (2019)

27. Heng, L., Choi, B., Cui, Z., et al.: Project AutoVision: localization and 3D scene perception for an autonomous vehicle with a multi-camera system. In: 2019 International Conference on Robotics and Automation (ICRA) (2019)

28. Christian Wolff, Dipl.-Ing. (FH): Radartutorial. https://www.radartutorial.eu/11.coherent/co06.en.html

29. Zhang, Y., Carballo, A., Yang, H., Takeda, K.: Perception and sensing for autonomous vehicles under adverse weather conditions: a survey. ISPRS J. Photogramm. Remote. Sens. **196**, 146–177 (2023)

30. GPS.gov: GPS overview. https://www.gps.gov/systems/gps/

31. Nagaoka, S.: Evaluation of attenuation of ultrasonic wave in air to measure concrete roughness using aerial ultrasonic sensor. Int. J. GEOMATE (2018)

32. Summon Your Tesla. https://www.tesla.com/blog/summon-your-tesla-your-phone
33. Javanmardi, E., Gu, Y., Javanmardi, M., Kamijo, S.: Autonomous vehicle self-localization based on abstract map and multi-channel LiDAR in urban area. IATSS Research (2019)
34. Choi, J.: Hybrid map-based SLAM using a Velodyne laser scanner. In: 17th International IEEE Conference on Intelligent Transportation Systems (ITSC) (2014)
35. Leonard, J.J., Durrant-Whyte, H.F.: Simultaneous map building and localization for an autonomous mobile robot. In: Proceedings IROS '91: IEEE/RSJ International Workshop on Intelligent Robots and Systems '91 (1991)
36. REV 7. https://ouster.com/blog/digital-lidar-realizing-the-power-of-moores-law
37. Yoneda, K., Suganuma, N., Yanase, R., Aldibaja, M.: Automated driving recognition technologies for adverse weather conditions. IATSS Research. **43**, 253–262 (2019)
38. Xu, Y., John, V., Mita, S., et al.: 3D point cloud map-based vehicle localization using stereo camera. In: 2017 IEEE Intelligent Vehicles Symposium (IV) (2017)
39. Carballo, A., Monrroy, A., Wong, D., et al.: Characterization of multiple 3D LiDARs for localization and mapping performance using the NDT algorithm. In: 2021 IEEE Intelligent Vehicles Symposium Workshops (IV Workshops) (2021)
40. Liu, W., et al.: SSD: Single Shot MultiBox Detector. In: Computer Vision – ECCV 2016, pp. 21–37 (2016)
41. Granström, K., Baum, M., Reuter, S.: Extended object tracking: introduction, overview, and applications. J. Adv. Inf. Fusion (2017)
42. Schulz, J., Hubmann, C., Lochner, J., Burschka, D.: Interaction-aware probabilistic behavior prediction in urban environments. In: 2018 IEEE/RSJ International Conference on Intelligent Robots and Systems (IROS) (2018)
43. Do, Q.H., Nejad, H.T.N., Yoneda, K., Ryohei, S., Mita, S.: Vehicle path planning with maximizing safe margin for driving using Lagrange multipliers. In: 2013 IEEE Intelligent Vehicles Symposium (IV) (2013)
44. Wojtanowski, J., Zygmunt, M., Kaszczuk, M., Mierczyk, Z., Muzal, M.: Comparison of 905 nm and 1550 nm semiconductor laser rangefinders' performance deterioration due to adverse environmental conditions. Opto-Electron. Rev. 22 (2014)
45. Zang, S., Ding, M., Smith, D., et al.: The impact of adverse weather conditions on autonomous vehicles: how rain, snow, fog, and hail affect the performance of a self-driving car. IEEE Vehicular Technology Magazine (2019)
46. Caccia, L., Hoof, H. van, Courville, A., Pineau, J.: Deep generative modeling of LiDAR Data. In: 2019 IEEE/RSJ International Conference on Intelligent Robots and Systems (IROS)
47. Gourova, R., Krasnov, O., Yarovoy, A.: Analysis of rain clutter detections in commercial 77 GHz automotive radar. In: 2017 European Radar Conference (EURAD) (2017)
48. Aldibaja, M., Suganuma, N., Yoneda, K.: Robust intensity-based localization method for autonomous driving on snow-wet road surface. IEEE Trans. Industr. Inf. **13**, 2369–2378 (2017)
49. OS1 Sensor. https://ouster.com/products/scanning-lidar/os1-sensor
50. LiDAR vs Camera: driving in the rain. https://ouster.com/blog/lidar-vs-camera-comparison-in-the-rain
51. Introducing the l2x Chip. https://ouster.com/blog/introducing-the-l2x-chip
52. LIBRE-dataset. https://sites.google.com/g.sp.m.is.nagoya-u.ac.jp/libre-dataset
53. Kocic, J., Jovicic, N., Drndarevic, V.: Sensors and sensor fusion in autonomous vehicles. In: 2018 26th Telecommunications Forum (TELFOR) (2018)
54. Zou, J., Zheng, H., Wang, F.: Real-Time target detection system for intelligent vehicles based on multi-source data fusion. Sensors **23**, 1823 (2023)
55. Ravindran, R., Santora, M.J., Jamali, M.M.: Multi-object detection and tracking, based on DNN, for autonomous vehicles: a review. IEEE Sens. J. (2021)

56. Singh, C.H., Mishra, V., Jain, K., Shukla, A.K.: FRCNN-based reinforcement learning for real-time vehicle detection, tracking and geolocation from UAS Drones (2022)
57. How Autonomous Vehicles Sensors Fusion Helps Avoid Deaths. https://intellias.com/sensor-fusion-autonomous-cars-helps-avoid-deaths-road/
58. Elfring, J., Appeldoorn, R., van den Dries, S., Kwakkernaat, M.: Effective world modeling: multi-sensor data fusion methodology for automated driving. Sensors (2016)
59. Kim, S., Song, W.-J., Kim, S.-H.: Double weight-based SAR and infrared sensor fusion for automatic ground target recognition with deep learning. Remote Sens. (2018)
60. Arnold, E., Al-Jarrah, O.Y., Dianati, M., Fallah, S., Oxtoby, D., Mouzakitis, A.: A survey on 3D object detection methods for autonomous driving applications. IEEE Trans. Intell. Transp. Syst. **20**, 3782–3795 (2019)
61. Liang, W., Xu, P., Guo, L., Bai, H., Zhou, Y., Chen, F.: A survey of 3D object detection. Multimedia Tools Appl. **80**, 29617–29641 (2021)
62. Qian, R., Lai, X., Li, X.: 3D object detection for autonomous driving: a survey. Pattern Recogn. **130**, 108796 (2022)
63. Rolling Zettabytes: quantifying the data impact of connected cars. https://www.datacenterfrontier.com/connected-cars/article/11429212/rolling-zettabytes-quantifying-the-data-impact-of-connected-cars
64. Qian, K., Zhu, S., Zhang, X., Li, L.E.: Robust multimodal vehicle detection in foggy weather using complementary Lidar and Radar signals. In: 2021 IEEE/CVF Conference on Computer Vision and Pattern Recognition (CVPR) (2021)
65. Mai, N.A.M., Duthon, P., Khoudour, L., Crouzil, A., Velastin, S.A.: 3D object detection with SLS-fusion network in foggy weather conditions. Sensors **21**, 6711 (2021)
66. Liu, Z., Cai, Y., Wang, H., et al.: Robust target recognition and tracking of self-driving cars with radar and camera information fusion under severe weather conditions. IEEE Trans. Intell. Transp. Syst. (2022)
67. John, V., Mita, S., Lakshmanan, A., Boyali, A., Thompson, S.: Deep visible and thermal camera-based optimal semantic segmentation using semantic forecasting. J. Auton. Veh. Syst., 1–10 (2021)
68. Rawashdeh, N.A., Bos, J.P., Abu-Alrub, N.J.: Drivable path detection using CNN sensor fusion for autonomous driving in the snow. In: Autonomous Systems: Sensors, Processing, and Security for Vehicles and Infrastructure 2021 (2021)
69. Vachmanus, S., Ravankar, A.A., Emaru, T., Kobayashi, Y.: Multi-modal sensor fusion-based semantic segmentation for snow driving scenarios. IEEE Sens. J. (2021)
70. Yang, B., Guo, R., Liang, M., Casas, S., Urtasun, R.: RadarNet: exploiting radar for robust perception of dynamic objects. In: European Conference on Computer Vision (2020)
71. Vertens, J., Zurn, J., Burgard, W.: HeatNet: bridging the day-night domain gap in semantic segmentation with thermal images. In: 2020 IEEE/RSJ International Conference on Intelligent Robots and Systems (IROS) (2020)
72. Michaelis, C., Mitzkus, B., Geirhos, R., et al.: Benchmarking robustness in object detection: autonomous driving when winter is coming. ArXiv, abs/1907.07484 (2019)

Semantic Segmentation Using Events and Combination of Events and Frames

M. Ghasemzadeh[(✉)] [iD] and S. B. Shouraki[(✉)] [iD]

Department of Electrical Engineering, Sharif University of Technology, Tehran, Iran
ghasemzadehmehdi07@gmail.com, bagheri-s@sharif.edu

Abstract. Event cameras are bio-inspired sensors. They have outstanding properties compared to frame-based cameras: high dynamic range (120 vs 60), low latency, and no motion blur. Event cameras are appropriate to use in challenging scenarios such as vision systems in self-driving cars and they have been used for high-level computer vision tasks such as semantic segmentation and depth estimation. In this work, we worked on semantic segmentation using an event camera for self-driving cars. i) This work introduces a new event-based semantic segmentation network and we evaluate our model on DDD17 dataset and Event-Scape dataset which was produced using Carla simulator. ii) Event-based networks are robust to lighting conditions but their accuracy is low compared to common frame-based networks, for boosting the accuracy we propose a novel event-frame-based semantic segmentation network that it uses both images and events. We also introduce a novel training method (blurring module), and results show our training method boosts the performance of the network in recognition of small and far objects, and also the network could work when images suffer from blurring.

Keywords: Event-Based Camera · Semantic Segmentation · Computer Vision · Self-Driving Cars · Sensor Fusion

1 Introduction

Events cameras are novel vision sensors, they work completely different from frame-based cameras. Event cameras focus on only changes in the scene and they record and return brightness changes per pixel instead of recording all the scene at regular intervals. In event cameras, each pixel works independently with microsecond resolution [17], and it generates a stream of events, and each event contains four numbers (x, y, t, p), x and y show pixels location, and t reveals when brightness change occurred and p shows the polarity of brightness change.

Properties of event cameras could motivate experts to develop new computer vision models for low and high-level tasks, and Event-Based models could work in extreme illumination (bright or dark) conditions, hence they are very appropriate to use in vision systems of self-driving cars. Event cameras could improve safety and reliability when

---

GitHub Page: https://github.com/mehdighasemzadeh/Event-Frame-Based-Semantic-Segmentation.git and https://github.com/mehdighasemzadeh/Event-Based-Semantic-Segmentation.git.

frame-based cameras suffer from low dynamic range, motion blur, and high latency. For example in Fig. 1 [6], captured images by a frame camera suffer from a lack of information, in this situation, event cameras are more reliable. On the other hand, the lack of available datasets for several tasks in event-based models causes difficulty in developing models. However event-based models have been used in various high-level tasks such as optical flow estimation [18, 19], reconstruction of visual information [20, 21], depth estimation [5, 22], and semantic segmentation [1, 10].

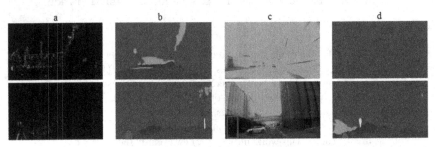

Fig. 1. Experiments in [6] showed frame-based models are not able to work well in extreme illumination (bright or dark) conditions compared to event-based models. (violet: street; green: vegetation; red: person; blue: car; yellow: object; gray: background). a: events, b: prediction from an Event-Based model, c: frame, and d: prediction from a Frame-Based model. (Color figure online)

Image semantic segmentation is a crucial task in autonomous driving, frame-based semantic segmentation models are accurate, but in challenging conditions, they are not able to work properly and the output of these models is not valid and reliable [1, 6], and Fig. 1. Event cameras are robust to lighting conditions, therefore event-based semantic segmentation models promise to improve robustness and reliability. These unique properties are a major motivation to employ event cameras in challenging tasks. In this paper, we introduce a new event-based semantic segmentation model which was inspired by SwiftNetRN-18 [23], they used two parallel encoders (pyramid-based architecture) and each encoder was built using pre-trained ResNet18 [24] on ImageNet. First, we evaluated our network on DDD17 (Alonso) dataset which was introduced by [1], and they released the first public dataset in this task and it became a benchmark but it is a pseudo-label dataset with only six classes. For demonstrating our model performance we needed a dataset with perfect labels and more classes, therefore, we trained and evaluated our networks using Event-Scape dataset [22] which was recorded in Carla simulator [11]. Our experiment shows our event-based segmentation is accurate enough in recognizing some classes which are generated a significant number of events for them, but in contrast for some classes such as road and sidewalk, intensities do not change by moving, hence a few numbers of events are generated, so the event-based model is not accurate for these classes as well as frame-based models. To address this challenge, we propose a new event-frame-based semantic segmentation, our main goal in this model is to combine the advantages of frame-based models and event-based models, events and images could complement each other perfectly if the model is robust and accurate. Our model uses both images and events and also our network is trained using a new method (blurring module in Sect. 3.2) for achieving robustness to blurred images. Experiments show our training method is efficient and the model achieves more accuracy and reliability compared to the event-based model also it has a robust performance when blurred images are applied to the network.

2 Related Work

The first event-based semantic segmentation was [1], the authors released the first public dataset using DDD17 [2] which was recorded by the DAVIS346B [3] and it contains grayscale images, events, and labels for autonomous driving such as steering angle but semantic segmentation labels do not exist on it, therefore labels were generated by a frame-based pre-trained network in [1]. They used an Xception-type network [4], and investigated the effect of event representation methods on the accuracy, and also demonstrated how event-based models have robust performance in challenging scenarios. However generated dataset in [1] has low resolution, pseudo labels, poor image quality, and only six classes. Therefore Event-Scape dataset [22] with higher resolution and quality could be a perfect alternative. In paper [6], authors improved the accuracy of [1] by synthetic events which are obtained from videos, they use real and synthetic events for training, which could improve the accuracy, but the proposed method requires video datasets which is not common for semantic segmentation tasks. For reducing dependency on videos, [7] proposed combining unlabeled events and frames from DAVIS with labeled image datasets such as Cityscape [8], they reported an improvement in accuracy and performance using transfer learning, but the proposed method needs paired event and image data. Another approach in [9] proposed event-to-image transfer to boost the performance of previous models. In [10], the authors proposed an unsupervised domain adaptation (UDA) method that leverages image datasets to train neural networks for event data.

In this work, we propose a straightforward event-based semantic segmentation network by inspired [23], our network shows improvements in accuracy and performance on DDD17 dataset, and we also train and evaluate our model on a new benchmark Event-Scape dataset which is a dataset with the perfect and exact label. Accurate detection in event-based models depends on objects' motion, for example, if a pedestrian or any object moves in the scene, event-based models are able to detect it, even for small objects but for some classes such as road and sidewalk classes, event cameras do not trigger, therefore event-based models suffer from detection of motionless objects. On the other hand, frame-based models have a perfect performance in good lighting conditions but they are not robust to lighting conditions and their safety and reliability are not as same as event-based models, it might seem event-based models and frame-based models could complement each other, therefore we present a new event-frame based network. In [1] they proposed an event-frame-based semantic segmentation model using an Xception-type network [4], the method for the fusion of events and frames is purely concatenation of events and frame and then employ it as input of the network, and their experiment showed combining events with frames could boost accuracy but the authors did not discuss robustness. In this work, we use an encoder-decoder architecture with two parallel encoders and 1×1 convolutions are used for the fusion of feature maps after each block, and a novel training method is used, the impact of that is shown in Fig. 8 and Fig. 12.

3 The Proposed Segmentation Models

Semantic segmentation is extensively used in autonomous driving, and it has a vital impact on these systems, and event-based semantic segmentation models have been used to boost reliability and safety using the unique properties of event cameras. We first introduce our event-based semantic segmentation network, and then we recommend a novel semantic segmentation model that it employs images and events for doing this task, we evaluate the proposed network regarding accuracy and robustness.

3.1 Event-Based Semantic Segmentation Model

We propose to use a pyramid structure for event-based semantic segmentation, our encoder-decoder network uses two parallel encoders and a light decoder. ResNet-18 [24] could be a good option to employ as a light and real-time encoder. Encoders are applied to event representation in different resolutions, H × W and H/2 × W/2 (pyramid resolution). Figure 2 shows them in blue. In the following, we present all parts of the proposed network.

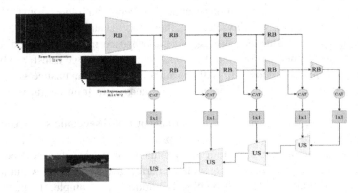

Fig. 2. The event-based network architecture, RB: ResNet-18 Block, CAT: concatenation, 1 × 1: 1 × 1 convolution, and US: upsampling block.

Encoder: We propose to use ResNet18 [24] because i) it is compatible with real-time models, ii) it is appropriate for fine-tuning, in this work, 5 × 5 convolution is used in the first layer instead of 7 × 7, maxpooling after the first layer is removed and ResNet Blocks (RB) have the same structure with original ResNet-18[24].

Decoder: UpSampling blocks (US) should be as simple as possible to increase speed and decrease the number of network parameters. Bilinear interpolation is used for upsampling of the low-resolution feature maps then they are mixed with same-size feature maps from lateral connections, this mixing is done by two convolution layers and each convolution is 3 × 3.

Fusion of Feature Maps: Our networks require a fusion of feature maps, our experiment shows concatenation of feature maps, and then combining them by 1×1 convolution is efficient to boost accuracy in the event-based semantic segmentation network, and also it is a good method for combining feature maps of events and images, and this method could boost accuracy and robustness in the event-frame-based semantic segmentation network.

Lateral Connection: In semantic segmentation networks, skip connection (lateral connection) is used to skip feature maps from encoder to decoder which could improve the reconstruction of spatial features. Lateral connection and the effect of that on accuracy are discussed in [23], their experiments revealed if lateral connections are taken before ReLU (Fig. 3), it could boost accuracy. We tried this issue in our network with event data, and we obtained the same results, it could boost the accuracy of our model.

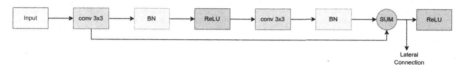

Fig. 3. Last unit in each Block. Conv 3×3: 3×3 convolution, BN: batch normalization, and SUM: summation.

3.2   Event-Frame-Based Semantic Segmentation Model

For improving the accuracy and performance of the event-based model, we propose to use both images and events. As previously mentioned, event cameras have several advantages over traditional cameras such as high dynamic range. These unique properties allow event-based networks to be able to work well in the toughest lighting condition indeed these networks are robust to lighting conditions. But with respect to event cameras, they generate data if brightness of pixels changes, this data format causes weakness in accuracy compared to frame-based networks. Combining the advantages of both networks is our goal, indeed robustness from an event camera, and high accuracy from a standard camera will be achieved. We propose a novel event-frame-based semantic segmentation, two parallel encoders are used, one of the encoders is applied to event representations and another one is applied to images, and then feature maps are combined by using concatenation and 1×1 convolution. The architecture of encoders, the decoder, and the fusion method are the same with 3.1 (Fig. 4).

Lateral Connection: We perused the consequences of taking lateral connections before and after ReLU, our experiments revealed if lateral connections are taken before ReLU in the last unit of each RS block, it improves accuracy between 3% and 4% compared to taking lateral connections after ReLU, but it has a negative effect on the performance of the network in challenging lighting conditions. Eventually, we selected taking lateral connection after ReLU to achieve more robustness to lighting conditions (Fig. 5).

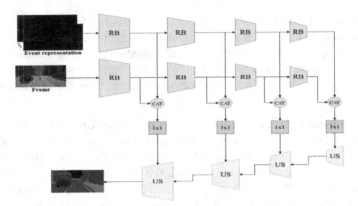

Fig. 4. The Event-Frame-based network architecture, RB: ResNet-18 Block, CAT: concatenation, 1×1: 1×1 convolution, and US: upsampling block.

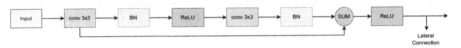

Fig. 5. Last unit in each Block. Conv 3×3: 3×3 convolution, BN: batch normalization, and SUM: summation.

Our Training Method: Figure 1 presented in [6], their experiment showed frame-based models are not able to work well in extreme illumination (bright or dark) conditions compared to event-based models, we combine events and images in our network and the main goal is combining advantages of both sensors (high accuracy using frames and robustness to lighting conditions using events) and there are a few samples that are similar to Fig. 1 in the DDD17 dataset [2] and there is no sample that is similar to Fig. 1 in Event-Scape dataset [22]. Our experiments showed if standard methods are used for training the network, events are just used to boost accuracy and improve the detection of some classes such as pedestrian class, and the expected robustness is not achieved. if we consider images which are taken by a frame camera in challenging lighting conditions, almost all of them suffer from a type of blurring, for example in Fig. 1, a pedestrian in a red rectangle is not revealed well, because of the low dynamic range of the frame camera, we could claim it is a kind of blurring. For producing more samples whose images suffer from low dynamic range, we propose the blurring module. A Gaussian filter in the blurring module is applied to images and the kernel size of the filter determines the degree of blurring. In our model, the blurring module was employed to acquire robustness to blurred images and also this method improves recognition of small and far objects Fig. 8.

Blurring Module: During training, with a probability of 50%, a Gaussian filter is applied to images and the kernel size is chosen randomly between 3 and 255. Then images are passed to Data Augmentation Module Fig. 6.

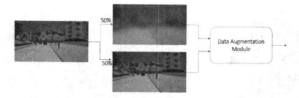

Fig. 6. Blurring Module is used for training the Event-Frame network.

4 Experiment

In Sects. 4.1 and 4.2, we evaluate the networks on DDD17 [1] and Event-Scape dataset [22] respectively, and also implementation details will be reported.

4.1 Our Networks on DDD17 Dataset

DDD17 [2] was recorded by DAVIS346B [3], in [1] several sequences of that were used for training and testing, and also authors in [1] generated semantic segmentation labels for these sequences using a pre-trained network and grayscale images and also It became a benchmark in this task. Resolution and quality in this dataset are low and the authors needed to merge multiple classes for reducing the granularity of the labels and these classes are: flat (road and non-road), background (sky and construction), object, vegetation, human, and vehicle. We trained and evaluated our model using it, in the following, training and implementation details will be reported.

Our Event Based Network[1]
Event Representation: In [1], authors used a six-channel event representation and these channels were filled by histograms of positive and negative events in each pixel [12, 13], temporal mean (M), and standard deviation (S) of positive and negative events in each pixel [1]. We also used these methods for event representation.

Loss Function: We used CrossEntropy loss and DICE loss [25], our network was trained with (CrossEntropy loss + DICE loss) in the first half of epochs and then just DICE loss was used.

Implementation Details: Adam optimizer with an initial learning rate of 5e−4 and an exponential learning rate decay schedule was used. During training crop, shift, and flip are performed in the data augmentation module. 80K iterations with batch size 16 was used for training.

Our Event-Frame-Based Network[2]
Event representation, Loss function, and Implementation details are the same as in the previous section. We evaluate our Event-Frame network in this part and also for demonstrating the impact of the blurring module, we trained our network with and

---

[1] More Tests: https://youtu.be/AL911t6QpBA.
[2] More Tests: https://youtu.be/o8nz3FxwzZg.

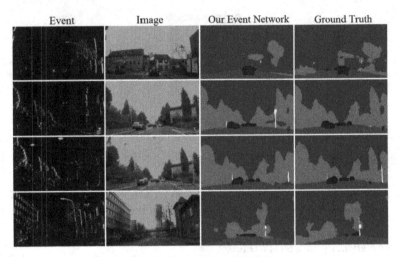

Fig. 7. Qualitative results on DDD17 dataset

without it. Our experiments show training the network using the blurring module achieves less MIoU approximately 1.7% compared to training the network without it Table 1, but this training method could boost the performance of the Event-Frame based network in recognition of small and far objects. (Fig. 8) and also improves reliability when frames suffer from blurring or low dynamic range.

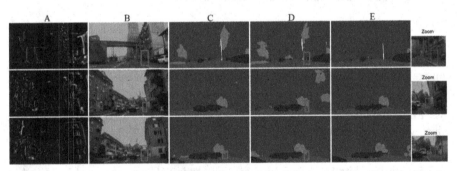

Fig. 8. Performance of our Event-Frame network (event + frame) with and without the blurring module on DDD17 dataset. A) Events, B) Images, C) training the network without the blurring module, D) training the network with the blurring module, E) Ground Truth.

4.2 Our Networks on Event-Scape Dataset

Event-Scape dataset [22] was recorded on Carla simulator [11], and it is a synthetic dataset but in comparison with DDD17 [1] has more classes, exact and perfect labels, and higher resolution therefore it would be a good alternative for DDD17 [1] and also it contains high-quality RGB images hence it is appropriate for using in the event-frame

| Events | Image | Our Event Network | Ground Truth |

Fig. 9. Qualitative results on DDD17 dataset: small objects in DDD17 dataset are sometimes missed, red rectangles show an example of this problem, low quality of DDD17 dataset and pseudo labels cause that. Therefore reported results (accuracy and MIoU) could not be precise and they could be a bit higher or lower.

Table 1. Results on DDD17 [1], Event-based and Event-Frame based models. Our event-frame network has achieved the highest MIoU and accuracy on DDD17 [1]. The accuracy of our event network is close to the highest reported accuracy (ESS [10]).

Model	Input Data	Accuracy	MIoU	Parameters (M)
EV-Segnet [1]	Event	89.76	54.81	29
EV-Segnet [1]	Event + Frame	95.22	68.36	29
Vid2E [6]	Event	–	45.48	–
EvDistill (2ch) [7]	Event	–	57.16	59
EvDistill (Mch) [7]	Event	–	58.02	59
DTL [9]	Event	–	58.80	60
ESS [10]	Event	91.08	**61.37**	12
ESS [10]	Event + Frame	90.37	60.43	12
HALSIE [16]	Event + Frame	92.50	60.66	1.82
Our Event network	Event	89.16	61.11	25
Our Event-Frame network without the blurring module	Event + Frame	**95.72**	**72.20**	26
Our Event-Frame network with the blurring module	Event + Frame	**94.76**	**70.50**	26

based network and comparing the event network and the event-frame network. We train and evaluate our networks using it in this section.

Our Event Based Network[3]

Dataset: Event-Scape dataset [22] contains 536, 103, and 119 sequences for training, validation, and test respectively. We trained our network using the training part and the results of the test part are reported. Image and event resolution in this dataset is 256 × 512 and this resolution is higher than DDD17 [1] but compared to common datasets such as Cityscape [8] is not enough, therefore, we needed to reduce the granularity of the labels so we fuse some classes and these classes are: background (sky and unlabeled),

---

[3] More Tests: https://youtu.be/Q1pNcZDNzos.

object (buildings, fences, walls), human, poles (poles, traffic Signs), road, sidewalk, vegetation, and vehicle. Event-Scape dataset [22] contains more than 171,000 samples, for decreasing overlap between samples we only use samples that are multiples of 7k in each sequence.

Event Representation: We propose to use histograms of positive and negative events in each pixel which is introduced by [12, 13], a temporal mean method which is proposed by [1] for positive and negative events, Recent Event Representation method [1, 14] for positive and negative events, and Event Volume method [15] with two channels (B = 2), as a result, our representation has 8 channels.

Implementation Details: Loss function in Sect. 4.1 was employed. Adam optimizer with an initial learning rate of 5e−4 and an exponential learning rate decay schedule is used. During training crop, shift and flip were performed in data augmentation module. 132K iterations with batch size 8 was used for training.

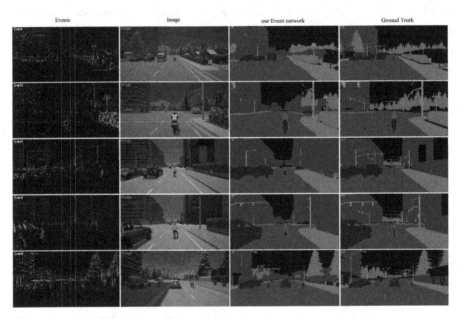

Fig. 10. Qualitative results of our Event network on Event-Scape dataset.

Our Event-Frame Network[4]

As previously mentioned, combining events and images could boost the performance of semantic segmentation networks and experiments in Sect. 4.2 using Event-Scape [22] dataset with more classes compared to DDD17 [1] showed the event-based network has good performance in recognizing objects which they have movement, even for small objects such as pedestrians. But these models suffer in recognizing objects with no

---

[4] More Tests: https://youtu.be/K6tkeT32Yi8.

movement, or generally when events are not produced for these classes such as road and sidewalk classes. To address this challenge we propose to use both images and events, and results show the network could be more reliable and accurate in recognizing all classes, moreover, the blurring module improves the robustness of the network to blurred images.

Dataset: The dataset in the previous section is used in this part but we are allowed to use RGB images and also road line is added to semantic segmentation classes, therefore, the dataset in this part includes 9 classes.

Event Representation and implementation details are the same as in the previous section.

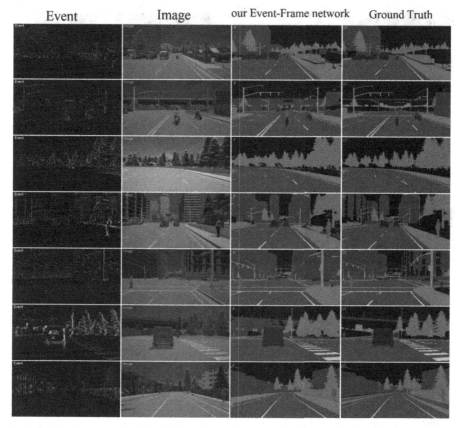

Fig. 11. Qualitative results of our Event-Frame network (Event + Frame) on Event-Scape dataset.

Robustness Test

Event-Scape dataset [22] contains high-quality RGB images and there is no failure such as blurring in images so for evaluating the robustness and performance of the model, a wide range of blurred images are fed to the network, and the network's predictions are shown in Fig. 12, results show our network is robust to a wide degree of blurred images although the accuracy decreases with increasing the blurring but outputs are still reliable. Furthermore, we evaluate our model when the event camera does not trigger any event, indeed a zero array is fed to the network instead of the event representation, and the output shows our model is still reliable (Fig. 13).

For revealing the robust performance of the network, all test images were blurred using a Gaussian filter with a kernel size of k, and then paired blurred images and events were fed to the network. In another experiment test images along with a zero array instead of event representation were fed to the network and the results of these experiments are reported in Table 3.

Fig. 12. Robustness test on Event-Scape dataset is done using a wide range of blurred images. Blurred images using a Gaussian filter with a kernel size of k are shown in the top row and predictions of the network with blurred images are shown in the bottom row. Results show our Event-frame network is robust to this problem, for example in the last column, the image is blurred with a kernel size of k = 253, but the output of the network is still reliable. This experiment shows our network and our training method are able to use both events and frames effectively.

Fig. 13. Robustness test when the event camera does not trigger any event (a zero array is applied to the network instead of the event representation.

Table 2. Results on Event-Scape dataset

Model	Input Data	Accuracy	MIoU	Parameters (M)
Our Event network	Event	84.96	52.74	25
Our Event-Frame network with blurring module	Event + Frame	90.05	65.23	26

For demonstrating the robust performance of the Event-Frame network, all test images are blurred by a Gaussian filter with a kernel size of k then they are fed to the network, and results are reported in Table 3. In another experiment (in the last row of Table 3), just images are applied to the network without event data (we assume that the event camera does not trigger any event).

Table 3. Robustness test[a] on Event-Scape dataset.

Model	Images are blurred with a kernel size of (k)	Using events	Accuracy	MIoU
Our Event-Frame network	Original Images	Yes	90.05	65.23
Our Event-Frame network	K = 11	Yes	88.70	62.06
Our Event-Frame network	K = 55	Yes	86.64	56.77
Our Event-Frame network	K = 111	Yes	85.57	54.50
Our Event-Frame network	K = 255	Yes	84.42	52.29
Our Event-Frame network	Original Images	No	82.25	53.90

[a]More Tests: https://youtu.be/EUNrJiVePPE

5 Conclusion

In this work, we introduced an event-based network for semantic segmentation and we evaluated our network using DDD17 and Event-Scape dataset our experiments showed the event-based network has good performance in recognizing some classes such as pedestrians and cars, but in recognizing some classes such as road is not accurate as well as frame-based models. For improving the performance of the event-based model, we proposed to use events along with images in an event-frame-based semantic segmentation network. Combining the advantages of event-based models and frame-based models was our goal for the event-frame-based network. Our experiments revealed if common training methods are used for training the event-frame-based network, events data are just used for boosting accuracy, and expected robustness to lighting conditions will not be obtained due to, we proposed to use the blurring module, and the experiments showed our method is efficient and the networks could be robust and accurate and also more reliable compared to the event-based model and frame-based model. Eventually combining events and frames could be beneficial if the model has advantages of both models (event-based model and frame-based model) namely high accuracy and robustness to lighting conditions, and our network has these advantages. In this work, we consider the most common failure in images indeed image blurring, for future works, other types of failure in images should be considered for achieving more effective robustness, and also other types of fusion methods and neural networks could be applied and evaluated.

References

1. Alonso, I., Murillo, A.C.: EV-SegNet: semantic segmentation for event-based cameras. In: IEEE Conference on Computer Vision and Pattern Recognition Workshops (CVPRW) (2019)
2. Binas, J., Neil, D., Liu, S.-C., Delbruck, T.: Ddd17: end-to-end Davis driving dataset. arXiv preprint arXiv:1711.01458 (2017)
3. Brandli, C., Berner, R., Yang, M., Liu, S.C., Delbruck, T.: A 240x180 130dB 3µs latency global shutter spatiotemporal vision sensor. IEEE J. Solid-State Circuits **49**(10), 2333–2341 (2014). https://doi.org/10.1109/JSSC.2014.2342715

4. Chollet, F.: Xception: deep learning with depthwise separable convolutions. In: IEEE Conference on Computer Vision and Pattern Recognition (CVPR), pp. 1800–1807 (2017). https://doi.org/10.1109/CVPR.2017.195

5. Gehrig, M., Aarents, W., Gehrig, D., Scaramuzza, D.: Dsec: a stereo event camera dataset for driving scenarios. IEEE Robot. Autom. Lett. (2021). https://doi.org/10.1109/LRA.2021.3068942

6. Gehrig, D., Gehrig, M., Hidalgo-Carrio, J., Scaramuzza, D.: Video to events: recycling video datasets for event cameras. In: Proceedings of the IEEE/CVF Conference on Computer Vision and Pattern Recognition, pp. 3586–3595 (2020)

7. Wang, L., Chae, Y., Yoon, S.H., Kim, T.K., Yoon, K.J.: Evdistill: asynchronous events to end-task learning via bidirectional reconstruction-guided cross-modal knowledge distillation. In: IEEE Conference on Computer Vision and Pattern Recognition (CVPR) (2021)

8. Cordts, M., et al.: The cityscapes dataset for semantic urban scene understanding. In: IEEE Conference on Computer Vision and Pattern Recognition (CVPR) (2016)

9. Wang, L., Chae, Y., Yoon, K.J.: Dual transfer learning for event-based end-task prediction via pluggable event to image translation. In: International Conference on Computer Vision (ICCV), pp. 2135–2145 (2021)

10. Sun, Z., Messikommer, N., Gehrig, D., Scaramuzza, D.: ESS: learning event-based semantic segmentation from stillimages. arXiv preprint arXiv:2203.10016 (2022)

11. Dosovitskiy, A., Ros, G., Codevilla, F., Lopez, A., Koltun, V.: CARLA: an open urban driving simulator. In: Conference on Robotics Learning (CoRL) (2017)

12. Moeys, D.P., et al.: Steering a predator robot using a mixed frame/event-driven convolutional neural network. In: 2016 Second International Conference on Event-Based Control, Communication, and Signal Processing (EBCCSP), pp. 1–8. IEEE (2016)

13. Maqueda, A.I., Loquercio, A., Gallego, G., García, N., Scaramuzza, D.: Event-based vision meets deep learning on steering prediction for self-driving cars. In: Proceedings of the IEEE Conference on Computer Vision and Pattern Recognition, pp. 5419–5427 (2018)

14. Lagorce, X., Orchard, G., Galluppi, F., Shi, B.E., Benosman, R.B.: Hots: a hierarchy of event-based time-surfaces for pattern recognition. IEEE Trans. Pattern Anal. Mach. Intell. 39(7), 1346–1359 (2017)

15. Zhu, A.Z., Yuan, L., Chaney, K., Daniilidis, K.: Unsupervised event based learning of optical flow, depth, and egomotion. In: Proceedings of the IEEE Conference on Computer Vision and Pattern Recognition, pp. 989–997 (2019)

16. Biswas, S.D., Kosta, A., Liyanagedera, C., Apolinario, M., Roy, K.: HALSIE – hybrid approach to learning segmentation by simultaneously exploiting image and event modalities. https://arxiv.org/abs/2211.10754 (2022)

17. Gallego, G., et al.: Event-based vision: a survey. IEEE Trans. Pattern Anal. Mach. Intell. (2020). https://doi.org/10.1109/TPAMI.2020.3008413

18. Bardow, P., Davison, A.J., Leutenegger, S.: Simultaneous optical flow and intensity estimation from an event camera. In: IEEE Conference on Computer Vision and Pattern Recognition (CVPR), pp. 884–892 (2016). https://doi.org/10.1109/CVPR.2016.102

19. Zhu, A.Z., Yuan, L., Chaney, K., Daniilidis, K.: Unsupervised event-based learning of optical flow, depth, and egomotion. In: IEEE Conference on Computer Vision and Pattern Recognition (CVPR) (2019)

20. Rebecq, H., Ranftl, R., Koltun, V., Scaramuzza, D.: High speed and high dynamic range video with an event camera. IEEE Trans. Pattern Anal. Mach. Intell. (2019). https://doi.org/10.1109/TPAMI.2019.2963386

21. Reinbacher, C., Graber, G., Pock, T.: Real-time intensity-image reconstruction for event cameras using manifold regularisation. In: British Machine Vision Conference (BMVC) (2016). https://doi.org/10.5244/C.30.9

22. Gehrig, D., Rüegg, M., Gehrig, M., Hidalgo-Carrio, J., Scaramuzza, D.: Combining events and frames using recurrent asynchronous multimodal networks for monocular depth prediction. IEEE Robot. Autom. Lett. (RA-L) (2021)
23. Oršić, M., Krešo, I., Bevandić, P., Šegvić,. S.: In defense of pre-trained imagenet architectures for real-time semantic segmentation of road-driving images. In: CVPR 2019 (2016)
24. He, K., Zhang, X., Ren, S., Sun, J.: Deep residual learning for image recognition. In: IEEE Conference on Computer Vision and Pattern Recognition (CVPR), pp. 770–778 (2016). https://doi.org/10.1109/cvpr.2016.90
25. Sudre, C.H., Li, W., Vercauteren, T., Ourselin, S., Jorge Cardoso, M.: Generalized Dice Overlap as a Deep Learning Loss Function for Highly Unbalanced Segmentations. In: Cardoso, M., et al. (eds.) DLMIA ML-CDS 2017. LNCS, vol. 10553, pp. 240–248. Springer, Cham (2017). https://doi.org/10.1007/978-3-319-67558-9_28

Deep Learning-Based Concrete Crack Detection Using YOLO Architecture

Elham Nabizadeh[1](✉) [ID] and Anant Parghi[2] [ID]

[1] Department of Civil Engineering, Hakim Sabzevari University, Sabzevar, Iran
ElhamNabizadeh20@gmail.com
[2] Department of Civil Engineering, Sardar Vallabhbhai National Institute of Technology,
Surat, India
amp@amd.svnit.ac.in

Abstract. Buildings, bridges and dams are important infrastructures which containing concrete; hence it is essential to understand how the concrete cracks when it is in service condition. The most common flaw in concrete structures is cracking, which reduces load-carrying capacity, stiffness, and durability. In this research, we employ deep learning methods to detect surface cracks in concrete buildings. The purpose of this research was to compare the detection capabilities of the YOLOv8 and YOLOv5 models. The models were quantitatively evaluated using evaluation measures like accuracy, recall, and mean average precision to analyze their detection performance. This study demonstrates that the YOLOv8 algorithm exhibits superior performance in detection accuracy compared to the YOLOv5 algorithms. Results show that the YOLOv8l model has the highest precision value, the YOLOv8x has the highest recall value, and the YOLOv8m and YOLOv8x have the highest mAP@50 value. Also, the mAP@50–90 values of these models are approximately equal and are the highest among other models.

Keywords: Concrete · Crack Detection · Deep learning · Computer vision · YOLO

1 Introduction

In infrastructure, concrete is the most commonly used man-made material. The cracks in concrete can damage the structure's stability since they change the stress field of each structure's element. The crack weakens the structure's bearing capacity dramatically and hastens its service life. For reinforced concrete structures, cracks can also cause steel reinforcement to corrode, thereby accelerating the cracking process. An inspection of concrete structures is conducted periodically to assess their overall condition and detect surface cracks or other damage. These types of manual inspections are expensive, time-consuming, and also need expensive equipment for the inspection of concrete structures. Hence, autonomous inspections can be used as alternatives to manual inspections of concrete structures.

Computer vision-based deep learning algorithms and image processing techniques have gradually replaced sensor-based crack detection techniques over the past few years

© The Author(s), under exclusive license to Springer Nature Switzerland AG 2023
M. Ghatee and S. M. Hashemi (Eds.): ICAISV 2023, CCIS 1883, pp. 182–193, 2023.
https://doi.org/10.1007/978-3-031-43763-2_11

[1]. Deep learning can overcome the limitations of previous approaches due to environmental sensitivity or image quality requirements, whereas image processing has essentially no drawbacks besides data dependency [2]. Deep learning models, such as convolutional neural network (CNN), were used by some researchers to address the lack of information in the literature about how supervised methods' performance is affected by the accessibility of training data [3, 4]. Cha et al. [5] used Faster R-CNN to detect cracks and four other forms of damage with an average accuracy of 87.8 percent. Wang et al. [6] used a convolutional neural network and principal component methods to detect pavement cracks, and their results showed a detection accuracy of over 90%. The YOLOv4-FPM model was used by Yu et al. [7] to resolve the issue of multi-scale fracture detection, and the findings show that it has a mAP of 0.976, or 0.064 more than YOLOv4. Cha et al. [8] suggested a deep learning model for crack detection utilizing a CNN, as well as a multiple damage detection technique that uses a CNN to identify steel corrosion, steel delamination, bolt corrosion, and concrete cracks.

Real-time crack identification was made possible using a deep-learning technique proposed by Choi et al. [9] that works on photos with shifting background data. Crack detection using R-CNN and a square-shaped indication for crack size estimation was proposed by Kim et al. [10]. Leakage monitoring systems could be employed in tunnels to save the hassle of relying on manual inspection alone. Xue et al. [11] used Mask-RCNN to detect seepage and compute accurate leakage areas. Mirbod and Shoar [12] proposed a smart concrete crack detection technique that extracts features using machine vision, artificial neural networks, and pattern recognition. Using convolutional neural networks, transfer learning, and a hybridized method, Yu et al. [13] suggested a vision-based automated system for surface condition identification of concrete photographs. An enhanced WGAN-GP network was proposed by Zhong et al. [14] to generate pavement picture datasets. In order to ensure the robustness of the proposed deeper WGAN-GP model, faster R-CNN, YOLOv3, and YOLOv4 models were trained on real crack images and used to produce crack images for region-level identification. In order to automatically detect cracks in a bridge, Li et al. [15] implemented a Faster R-CNN technique using VGG16 transfer learning. The F1-score that can be attained with their method is 94.10%. In order to provide extremely accurate and automatic semantic segmentation for a wide variety of fractures in engineered cementitious composites, Hao et al. [16] suggested a deep-learning technique. Long et al. [17] developed a method for measuring the rate of expansion of fracture length using deep learning as the primary data collection technique.

Qiu and Lau [18] built a drone equipped with the YOLO algorithm to monitor for pavement damage in real time. To improve detection accuracy and speed, various network architectures were reframed and compared. Zhang et al. [19] proposed an improved method for automatically detecting and classifying bridge surface cracks based on YOLO. In order to detect wind-erosion damage on concrete, Cui et al. [20] employed a transformer theory-based methodology to enhance YOLO-v4 and create an object identification method called MHSA-YOLOv4. Wu et al. [21] developed an enhanced YOLOv4 model that distinguishes concrete cracks from numerous misleading targets, based on the pruning method and the EvoNorm-S0 structure. Villanueva et al. [22] used the Yolov3 model and also developed an android application to detect cracks in reinforced concrete structures and classify them as being moderate, severe, or extremely serious. Zhao et al. [23] proposed a new image analysis-based crack feature pyramid network called Crack-FPN and YOLOv5 for crack region detection. They found that Crack-FPN and YOLOv5 techniques could retrieve features in a unique way and is easy to run. Dung et al. [24] used three existing CNN architectures in their study, such as VGG16, Inception-V3, and ResNet. Bang et al. [25] found road cracks using VGG-16 and ResNet-50 as encoders in transfer learning models based on the ImageNet dataset. Kim et al. [26] used a pre-trained AlexNet whose last layer was changed to include the five classes of crack, multiple lines, single line, complete surface, and plant. Zhang et al. [27] used fully convolutional neural network (FCN) with expanded convolution to figure out where cracks were in concrete. In their study, they used a residual network and dilated convolutions with different dilation rates to get maps of different receptive fields. Dorafshan et al. [28] did a comparison of edge-based algorithms and deep learning methods to see how well they could find cracks. Ali et al. [29] used a different form of the UNet model to find cracks on the surface of rocks. Crack identification on concrete structures was studied by Li et al. [30], who employed a linear support vector machine (SVM) with a greedy search strategy to eliminate background noise. Wang et al. [31] employed structured random forests to detect cracks in steel beams. Using a combination of the structured random forests method and the anti-symmetrical biorthogonal wavelets semi-reconstruction technique, they were able to locate the fracture edges. Combining fractal dimension and the UHK-Net network, as proposed by An et al. [32], is a novel approach to semantic recognition of concrete fractures.

This research evaluates the applicability of the single-stage detector YOLOv8 for detecting cracks in concrete surfaces and seeks to enhance its efficiency in terms of detection accuracy. This study also presents a methodology for automatic concrete crack segmentation using YOLOv7 and YOLOv5.The effectiveness of the different YOLOv8 versions is validated and tested. Here is how the rest of the report is structured. The YOLO algorithm is introduced in Sect. 2. In Sect. 3, we go over the methods for creating datasets and the functionality of the network. In Sect. 5, we give the segmentation and testing results for the concrete cracks, and in Sect. 6, we draw the necessary conclusions.

2   YOLO Network

While other object detection algorithms use a "pipe-line" of region proposal, classification, and duplication elimination, YOLO depicts object detection as a single regression challenge. Although YOLO can be used in many other frameworks, the most common one is Darknet for versions of YOLO lower than 5. In YOLO algorithms, images are downscaled before being fed into a single CNN, which then provides detection results based on the model's confidence threshold. While YOLO's first iteration had an optimized sum of square errors, it was not as accurate as more recent object identification models. To improve the resiliency of the model in different scenarios, the YOLO method uses online data augmentation to supplement the input data with additional photos. YOLO models were utilized to improve YOLO's performance in several contexts where quick detection was required. In May of 2020, a new developer, Ultralytics, released YOLOv5, and while there was some debate over whether or not it counted as a new version of YOLO, the deep learning community ultimately acknowledged it as such.

YOLOv5's native framework, PyTorch, allows for quicker training times. Available in four different sizes (s, m, l, and x), YOLOv5 provides fast detection with the same accuracy as YOLOv4 in terms of performance metrics. The YOLOv5 architecture is shown in Fig. 1. As can be seen, the YOLOv5 network consists of an object-recognition output node, a feature-fusion neck node, and a feature-extraction backbone node.

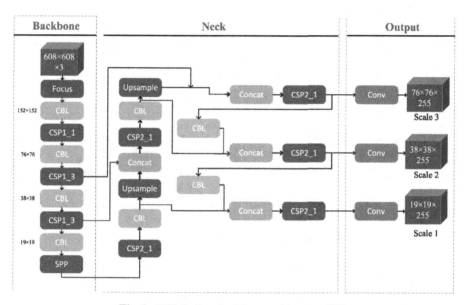

Fig. 1. YOLOv5 methodology architecture [33].

The main network is a convolutional neural network, which uses several convolutions and pooling to derive feature maps of varying sizes from the input image. Figure 1 shows the four-layer feature maps used to build the core network. The neck network uses the feature maps of various sizes to combine the feature maps of various layers in order to obtain more contextual information and lessen data loss. The fusion process makes use of the pyramidal structures that are a hallmark of both FPN and PAN. The FPN structure allows for the transmission of powerful semantic features from the top feature maps to the lower feature maps. The focus module of the design concatenates and slices images for more efficient feature extraction during down sampling. The CBL component includes the convolution, normalizing, and Leaky relu activation function modules. Two distinct varieties of cross-stage partial networks (CSP) are supported in YOLOv5. CSP networks use cross-layer connectivity to speed up inference while retaining accuracy by shrinking the size of the model. In the backbone, a CSP network is made up of one or more residual units, whereas in the neck, CBL modules take the place of the residual units. The SPP module is also known as the spatial pyramid pooling module, and it is responsible for performing maximum pooling with a wide range of kernel sizes and then concatenating the features to fuse them. Pooling mimics the human visual system by representing image data at a higher level of abstraction through the use of dimensionality reduction techniques. The main cause is the compressed input feature map. It compresses the feature map and isolates the most important characteristics, lowering the network's computational burden [33].

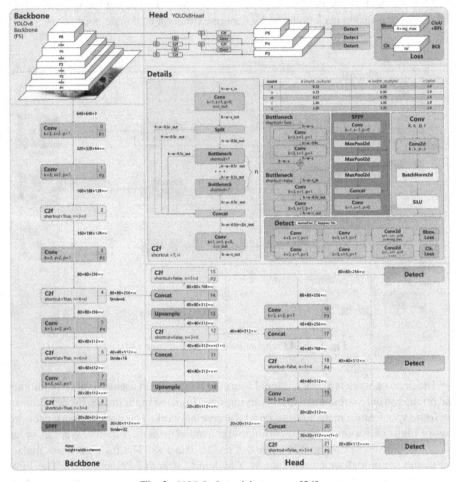

Fig. 2. YOLOv8 model structure [34].

The latest version of Ultralytics YOLO object detector and image segmentation system is called YOLOv8. YOLOv8 offers numerous advantages besides its versatility, making it an attractive choice for a variety of object recognition and image segmentation tasks. These features include a novel anchor-free detecting head, a brand-new backbone network, and a distinct loss function. YOLOv8 improves the model's capacity to capture high-level features by utilizing an updated backbone network built on EfficientNet. When compared to earlier versions of YOLO, accuracy and speed have improved in YOLOv8. YOLOv8 is a model that does not rely on anchors. This implies that it forecasts the center of an object directly rather than the offset from a known anchor box. The structure of the YOLOv8 model is shown in Fig. 2.

3   Dataset Preparation and Evaluation Metrics

The dataset has a crucial role in the experiment's success. To train the deep learning models for crack detection, a public available dataset was used from the robflow website [35]. For the crack detection dataset, 400 real images were used for the training set and 120 images for the test set and the labels of each image were stored in a text file with four coordinates that defined a rectangle box.

Precision, Recall, F-score and mAP are used to evaluate the training model. According to Eq. (1), the F1-score is the harmonic mean of the model's test accuracy, recall, and precision. The F1 score can range from a minimum of 0 (zero), which denotes either zero precision or zero recall, to a maximum value of 1, which denotes perfect precision and memory. The mAP is determined by taking the mean of the average precision for all classes, as indicated in Eq. (4), where q is the total number of queries and AveP(q) is the mean precision for each query. Therefore, mAP, a query-by-query average precision metric, can be computed to evaluate the efficacy of ML methods. The accuracy of a model improves as its Precision, Recall, and F1 metrics rise. Precision measures how many instances of a class were correctly predicted, while Recall measures how many instances of the class were correctly predicted. When Precision, Recall, and F1 are all high, the model's accuracy improves. Number of positive pixels and ground-truth value are defined in Eqs. (1) and (2), respectively, for true positives. The number of pixels assigned a positive score when the ground truth value is negative is called the false positive rate (FP), whereas the number of pixels assigned a negative score when the ground truth value is positive is called the false negative rate (FN).

$$Precision = \frac{TruePositive}{TruePositive + FalsePositive} \tag{1}$$

$$Recall = \frac{TruePositive}{TruePositive + FalseNegative} \tag{2}$$

$$F1 = \frac{2(Precision \times Recall)}{(Precision + Recall)} \tag{3}$$

$$mAP = \sum_{q=1}^{Q} \frac{AveP(q)}{Q} \tag{4}$$

4   Experimental Setup

The model was trained using a PC connected to Google Colab, a GPU Tesla T4, and the deep learning framework PyTorch 1.13. Input images were 500×500 pixels, the learning rate was 0.01, and the stochastic gradient descent (SGD) optimizer had 0.937 momentum. Table 1 displays the Experimental setup for training the model.

Table 1. Experimental setup.

Training parameters	Details
Batch-Size	16
Deep learning framework for training	PyTorch 1.11
Epochs	150
Optimization algorithm	SGD
img-size (pixels)	500×500
Optimizer momentum	0.937

5   Results and Discussion

Several versions of the YOLO algorithm including YOLOv8s, YOLOv8m, YOLOv8n, YOLOv8x, YOLO8l, YOLOv5s, and YOLOv5m were used to evaluate this research. For evaluating the performance of the YOLOv8s model, Fig. 3 displays the detection effect on some images randomly from the test set.

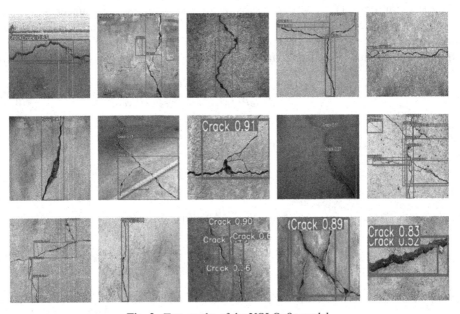

Fig. 3. Test results of the YOLOv8s model.

Five evaluation metrics are used to assess the models in terms of the detection effect. Figure 4 shows the precision, recall, mAP@50, mAP@50–90, P-R curves, and F1-scores for models. According to the depicted representation, Results show the performances of YOLOv8 versions under different evaluation metrics are good and it is evident that the YOLOv8 algorithm demonstrates an enhancement in comparison to YOLOv5.

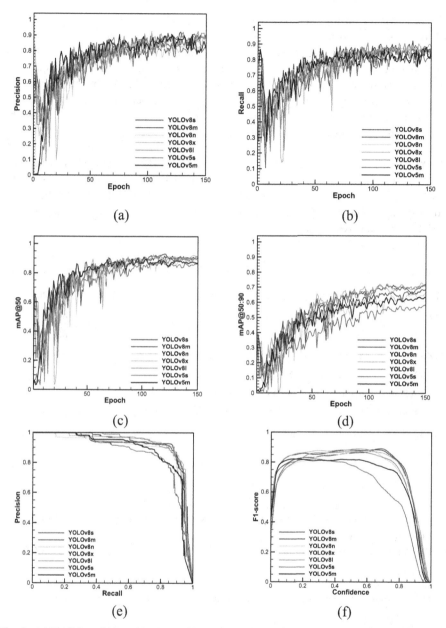

Fig. 4. (a) Precision, (b) Recall, (c) mAP@50, (d) mAP@50–90, (e) P-R curves, and (f) F1-scores for YOLO models.

There were 150 training epochs for the object detection models. Figure 4 (a) and (b) indicate an increase in precision and recall for the training models at around 50 epochs, followed by a trend toward stabilizing. Figure 4 (a) and (b) demonstrate that YOLOv5s and YOLOv5m are less accurate and less reliable than alternative models.

The mAP for an IoU of 0.5 is depicted in Fig. 4 (c) and the mAP for an IoU of 0.5 to 0.9 is depicted in of Fig. 4 (d). Figure 4 demonstrates that, within the first 50 epochs, accuracy steadily increases as the epoch number increases. Then, when the epoch number increased, there was some variation, followed by stability. Compared to YOLOv5, the mAP of the YOLOv8 algorithm is greater (Fig. 4). The trained model's accuracy vs recall is depicted in Fig. 4 (e). The relationship between a detection model's precision and recall is shown by the shape of its precision-recall curve. Area under the PR curve analysis was used to compute the mAP@50. As a result, reaching the upper right corner of the performance reward curve is where maximum efficiency is found. The YOLO models' F1-score curves are shown in Fig. 4 (f). According to the findings, the YOLOv5m and YOLOv5x models have lower F1 score values than the rest of the models. The research shows that YOLOv8 models are clustered in the upper right corner, indicating a high degree of accuracy and recall. The effectiveness of various detection models is summarized in Table 2.

Table 2. Experimental results.

Model	Precision	Recall	mAP@50	mAp@50–90	F1-score
YOLOv8s	90.19	87.32	90.56	68.68	88.73
YOLOv8m	90.36	85.91	91.31	71.94	88.07
YOLOv8n	87.30	85.21	88.65	65.43	86.24
YOLOv8x	90.38	88.02	91.49	71.18	89.18
YOLOv8l	91.50	83.48	89.59	70.74	87.30
YOLOv5s	82.58	81.69	86.18	58.37	82.13
YOLOv5m	80.94	86.73	87.29	63.30	83.26

The following findings are derived from the Table 2:

- The YOLOv8l model has the highest precision values of 1.31%, 1.14%, 4.2%, 1.12%, 8.92% and 10.56% higher than the YOLOv8s, YOLOv8m, YOLOv8n, YOLOv8x, YOLOv5s and YOLOv5m networks, respectively.
- The YOLOv8x has the highest recall values of 0.7%, 2.11%, 2.81%, 4.54%,6.33%, and 1.29% higher than those of the YOLOv8s, YOLOv8m, YOLOv8n, and YOLOv8l networks respectively.
- The mAP@50–90 values of The YOLOv8m and YOLOv8x are approximately equal and are the highest among other models.
- The F-1 score value from YOLOv8x is the highest value when compared to other models. The second-highest F1-score value is YOLOv8s.

6   Conclusions

Concrete is widely used as a building material in structural engineering. Consequently, it is essential to carry out a smart identification of the concrete erosion region. This study employed a deep learning approach to find cracks in preexisting concrete buildings. For this purpose, we trained YOLOv8 and YOLOv5 variants for crack identification in concrete. Several metrics, such as precision, recall, and mAP@50, were used to compare the effectiveness of the proposed models. Findings show that the YOLOv8 algorithm outperforms YOLOv5 in terms of precision and accuracy. YOLOv8 variants demonstrated promising performance and it can be concluded that YOLOv8 variants like YOLOv8s and YOLOv8m can be employed for a reliable concrete crack detection approach. The highest mAP@0.5 values were obtained by the YOLOv8m and YOLOv8x.

Acknowledgements. Without the central library of Sardar Vallabhbhai National Institute of Technology (SVNIT), we would have had a much tougher job referencing reliable sources for this research. Thus, the authors are grateful to the SVNIT library in Surat, India, for making the online database available to authors. The authors would also like to express their appreciation to the anonymous reviewers whose comments and suggestions helped to improve the quality of the article.

References

1. Bang, S., Baek, F., Park, S., Kim, W., Kim, H.: Image augmentation to improve construction resource detection using generative adversarial networks, cut-and-paste, and image transformation techniques. Autom. Constr. **115**, 103198 (2020). https://doi.org/10.1016/j.autcon.2020.103198
2. Yu, Z., Shen, Y., Sun, Z., Chen, J., Gang, W.: Cracklab: a high-precision and efficient concrete crack segmentation and quantification network. Dev. Built. Environ. **12**, 100088 (2022). https://doi.org/10.1016/j.dibe.2022.100088
3. Cheraghzade, M., Roohi, M.: Deep learning for seismic structural monitoring by accounting for mechanics-based model uncertainty. J. Build. Eng. **57**, 104837 (2022). https://doi.org/10.1016/j.jobe.2022.104837
4. Cheraghzade, M., Roohi, M.: Incorporating Uncertainty in Mechanics-based Synthetic Data Generation for Deep Learning-based Structural Monitoring International Institution of Earthquake Engineering and Seismology, Structural Research Center (2023)
5. Cha, Y.J., Choi, W., Suh, G., Mahmoudkhani, S., Büyüköztürk, O.: Autonomous structural visual inspection using region-based deep learning for detecting multiple damage types. Comput. Civ. Infrastruct. Eng. **33**, 731–747 (2018). https://doi.org/10.1111/mice.12334
6. Wang, X., Hu, Z.: Grid-based pavement crack analysis using deep learning. In: 2017 4th International Conference on transportation information and safety, pp. 917–924 (2017)
7. Yu, Z., Shen, Y., Shen, C.: A real-time detection approach for bridge cracks based on YOLOv4-FPM. Autom. Constr. **122**, 103514 (2021). https://doi.org/10.1016/j.autcon.2020.103514
8. Cha, Y.J., Choi, W., Büyüköztürk, O.: Deep learning-based crack damage detection using convolutional neural networks. Comput. Civ. Infrastruct. Eng. **32**, 361–378 (2017). https://doi.org/10.1111/mice.12263
9. Choi, W., Cha, Y.: SDDNet: real-time crack segmentation. IEEE Trans. Ind. Electron. **67**, 8016–8025 (2020)
10. Kim, I.-H., Jeon, H., Baek, S.-C., Hong, W.-H., Jung, H.-J.: Application of crack identification techniques for an aging concrete bridge inspection using an unmanned aerial vehicle. Sensors **18**, 1881 (2018). https://doi.org/10.3390/s18061881
11. Xue, Y., Cai, X., Shadabfar, M., Shao, H., Zhang, S.: Deep learning-based automatic recognition of water leakage area in shield tunnel lining. Tunn. Undergr. Sp. Technol. **104**, 103524 (2020). https://doi.org/10.1016/j.tust.2020.103524

12. Mirbod, M., Shoar, M.: Intelligent concrete surface cracks detection using computer vision, pattern recognition, and artificial neural networks. Procedia Comput. Sci. **217**, 52–61 (2023). https://doi.org/10.1016/j.procs.2022.12.201

13. Yu, Y., Samali, B., Rashidi, M., Mohammadi, M., Nguyen, T.N., Zhang, G.: Vision-based concrete crack detection using a hybrid framework considering noise effect. J. Build. Eng. **61**, 105246 (2022). https://doi.org/10.1016/j.jobe.2022.105246

14. Zhong, J., et al.: A deeper generative adversarial network for grooved cement concrete pavement crack detection. Eng. Appl. Artif. Intell. **119**, 105808 (2023). https://doi.org/10.1016/j.engappai.2022.105808

15. Li, R., Yu, J., Li, F., Yang, R., Wang, Y., Peng, Z.: Automatic bridge crack detection using Unmanned aerial vehicle and Faster R-CNN. Constr. Build. Mater. **362**, 129659 (2023). https://doi.org/10.1016/j.conbuildmat.2022.129659

16. Hao, Z., Lu, C., Li, Z.: Highly accurate and automatic semantic segmentation of multiple cracks in engineered cementitious composites (ECC) under dual pre-modification deep-learning strategy. Cem. Concr. Res. **165**, 107066 (2023). https://doi.org/10.1016/j.cemconres.2022.107066

17. Long, X., Yu, M., Liao, W., Jiang, C.: A deep learning-based fatigue crack growth rate measurement method using mobile phones. Int. J. Fatigue **167**, 107327 (2023). https://doi.org/10.1016/j.ijfatigue.2022.107327

18. Qiu, Q., Lau, D.: Real-time detection of cracks in tiled sidewalks using YOLO-based method applied to unmanned aerial vehicle (UAV) images. Autom. Constr. **147**, 104745 (2023). https://doi.org/10.1016/j.autcon.2023.104745

19. Zhang, J., Qian, S., Tan, C.: Automated bridge surface crack detection and segmentation using computer vision-based deep learning model. Eng. Appl. Artif. Intell. **115**, 105225 (2022). https://doi.org/10.1016/j.engappai.2022.105225

20. Cui, X., Wang, Q., Li, S., Dai, J., Xie, C., Duan, Y., et al.: Deep learning for intelligent identification of concrete wind-erosion damage. Autom. Constr. **141**, 104427 (2022). https://doi.org/10.1016/j.autcon.2022.104427

21. Wu, P., Liu, A., Fu, J., Ye, X., Zhao, Y.: Autonomous surface crack identification of concrete structures based on an improved one-stage object detection algorithm. Eng. Struct. **272**, 114962 (2022). https://doi.org/10.1016/j.engstruct.2022.114962

22. Villanueva, A., et al.: Crack detection and classification for reinforced concrete structures using deep learning. In: 2022 2nd International Conference on Intelligent Technologies, p. 1–6 (2022). https://doi.org/10.1109/CONIT55038.2022.9848129

23. Zhao, W., Liu, Y., Zhang, J., Shao, Y., Shu, J.: Automatic pixel-level crack detection and evaluation of concrete structures using deep learning. Struct. Control Heal. Monit. **29**, e2981 (2022). https://doi.org/10.1002/stc.2981

24. Dung, C.V., Anh, L.D.: Autonomous concrete crack detection using deep fully convolutional neural network. Autom. Constr. **99**, 52–58 (2019). https://doi.org/10.1016/j.autcon.2018.11.028

25. Bang, S., Park, S., Kim, H., Kim, H.: Encoder–decoder network for pixel-level road crack detection in black-box images. Comput. Civ. Infrastruct. Eng. **34**, 713–27 (2019). https://doi.org/10.1111/mice.12440

26. Kim, B., Cho, S.: Automated vision-based detection of cracks on concrete surfaces using a deep learning technique. Sensors **18**, 3452 (2018). https://doi.org/10.3390/s18103452

27. Zhang, J., Lu, C., Wang, J., Wang, L., Yue, X.-G.: Concrete cracks detection based on FCN with dilated convolution. Appl. Sci. **9** (2019). https://doi.org/10.3390/app9132686

28. Dorafshan, S., Thomas, R.J., Maguire, M.: Comparison of deep convolutional neural networks and edge detectors for image-based crack detection in concrete. Constr. Build. Mater. **186**, 1031–1045 (2018). https://doi.org/10.1016/j.conbuildmat.2018.08.011

29. Ali. R., Chuah, J.H., Talip, M.S.A., Mokhtar, N., Shoaib, M.A.: Automatic pixel-level crack segmentation in images using fully convolutional neural network based on residual blocks and pixel local weights. Eng. Appl. Artif. Intell. **104**, 104391 (2021). https://doi.org/10.1016/j.engappai.2021.104391

30. Li, G., Zhao, X., Du, K., Ru, F., Zhang, Y.: Recognition and evaluation of bridge cracks with modified active contour model and greedy search-based support vector machine. Autom. Constr. **78**, 51–61 (2017). https://doi.org/10.1016/j.autcon.2017.01.019

31. Wang, S., Liu, X., Yang, T., Wu, X.: Panoramic crack detection for steel beam based on structured random forests. IEEE Access **6**, 16432–16444 (2018). https://doi.org/10.1109/ACCESS.2018.2812141

32. An, Q., et al.: Segmentation of concrete cracks by using fractal dimension and UHK-net. Fract. Fract. **6** (2022). https://doi.org/10.3390/fractalfract6020095

33. Zhu, L., Geng, X., Li, Z., Liu, C.: Improving yolov5 with attention mechanism for detecting boulders from planetary images. Remote Sens. **13**, 1–19 (2021). https://doi.org/10.3390/rs13183776

34. GitHub Actions Automate your workflow from idea to production (2023). https://github.com/RangeKing. Accessed 5 Apr 2023

35. Mauriello, N.: CrackDetect Dataset. Roboflow Universe (2021)

Generating Control Command
for an Autonomous Vehicle Based
on Environmental Information

Fatemeh Azizabadi Farahani[✉], Saeed Bagheri Shouraki, and Zahra Dastjerdi

Electrical Engineering Department, Sharif University of Technology, Tehran, Iran
Fatemeh00farahani@gmail.com

Abstract. This paper presents a novel CNN architecture using an end-to-end learning technique to predict the steering angle for self-driving cars. The front camera is the only sensor used to generate this control command. This network was trained and tested on Sully Chen Public Dataset, which contains image frames and steering angle data for each image. The test outcomes demonstrated that this model could generate a reasonably accurate steering angle for autonomous vehicles and perform about 60% better than the networks designed from 2017 to 2021. In addition, the problem of overfitting in previous networks has mainly been addressed with the help of the new network architecture and different data preprocessing.

Keywords: Self-Driving Cars · Deep Learning · Steering Angle

1 Introduction

In the past decade, advances in computer vision [1], robotics [2], and natural language processing (NLP) [3] have been primarily attributed to Deep Learning and Artificial Intelligence (AI). They also significantly influence the current academic and industrial revolution in autonomous driving. Convolutional and recurrent neural networks [4, 5], generative adversarial networks (GAN) [6], and reinforcement learning [7] are key components of neural networks that are used to generate commands for autonomous cars. Generating control commands such as steering angle is essential for self-driving cars. There are two types of command-producing strategies used in self-driving cars [8]. The modular perception-planning-action pipeline, which consists of a number of modules and is developed using deep learning techniques, is the first method. Combining these modules could result in control commands. The second approach is end-to-end systems which directly try to find the connections between sensory information and control commands. The perception-planning-action pipeline has several modules and is shown in Fig. 1. Perception [9] helps the car see the world around itself and recognize and classify the things it sees [10]. The prediction module creates an n number of possible actions or moves based on the environment. The decision-making module then chooses the best action among them.

M. Ghatee and S. M. Hashemi (Eds.): ICAISV 2023, CCIS 1883, pp. 194–204, 2023.
https://doi.org/10.1007/978-3-031-43763-2_12

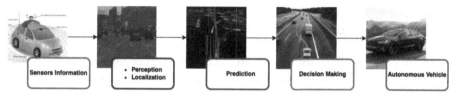

Fig. 1. The perception-planning-action pipeline approach for producing driving commands in self-driving cars

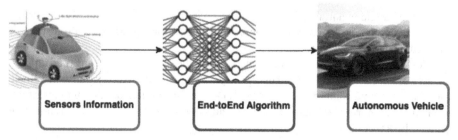

Fig. 2. The end-to-end approach for producing driving commands in self-driving cars

The goal of end-to-end learning (see Fig. 2) is that cars learn to drive just like humans do; by watching humans drive. In other words, they use end-to-end learning, a machine learning method, to directly map the correlation between variables. The pipelining technique divides the work of generating steering angle into three components: lane identification [11], finding a geometric path [12, 13], and controls logic [14, 15]. The other approach generates a steering angle directly by finding the variables' relationships.

The pipeline strategy is expensive and ineffective since it requires additional configuration to create an integrated modular system [16]. As a lot of parameter tweaking is done manually throughout this integration process, it can be quite time-consuming and result in information loss. Moreover, errors may carry over from one processing stage to the next, resulting in erroneous results in the end.

To address these problems, a significant body of research has been dedicated to employing deep learning techniques. A prevailing strategy involves training a single model in an end-to-end and multi-task fashion [17–19]. By employing this technique, the model can be trained to utilize observation data collected from a group of sensors, thereby determining the appropriate final action. The model uses the extracted characteristics independently because human tuning is no longer required. As there are no explicit criteria involved, the end-to-end model autonomously optimizes itself by learning directly from the training data. Better performance and fewer human interventions are two key benefits of end-to-end learning.

An ongoing obstacle for end-to-end models is the task of efficiently encoding and extracting pertinent information, which is crucial for the controller module to translate into appropriate vehicular commands. Therefore, both meticulous data preparation and a well-designed network architecture play pivotal roles. In this research, we introduce a novel neural network architecture that embraces an end-to-end approach. Furthermore, we emphasize the importance of proper data preprocessing techniques to precisely predict steering angles for self-driving cars.

The report's remaining sections are arranged as follows. Following a declaration of related work in Sect. 2, Sect. 3 details our implementation, including data preprocessing and CNN architecture, and Sects. 4 and 5 offer evaluation results and conclusions.

2 Related Works

The steering angle in autonomous cars has been calculated using various deep learning techniques, including the end-to-end approach. The following introduces four models that used this approach to predict the steering angle.

PilotNet: [20] proposed a network that generates the steering angle using road images. The network was trained and tested using images sampled from video from an in-vehicle front-facing camera combined with appropriate steering commands. The network architecture includes normalization, convolution, and fully connected layers. This architecture can mainly detect objects that affect the steering angle.

J-net: This network [21] was made by modifying the Alexnet architecture [1]. The network also gets the raw signal (camera images) as an input, and the output is the steering angle. This network is built based on convolution and max-pooling layers. Compared to PilotNet, it has better results in terms of predictive power, and also, the architecture is less complex.

The two mentioned approaches solved the problem with the help of convolutional layers. These layers can extract features by convolving the input with filters. This ability makes them useful in many applications, such as image classification [22]. CNNs, on the other hand, make predictions one frame at a time and it is relatively poor at explaining the long-term behavior of the environment. A possible solution for assuming transient connections is the use of Long Short Term Memory (LSTM) [23]. Therefore, to overcome the limitations of CNN and take advantage of LSTM, in 2015, the Convolutional LSTM or Conv-LSTM was proposed [24], which can simultaneously model Spatiotemporal structures by converting spatial information into tensors. However, the drawbacks of Conv-LSTM are high computational costs and high memory consumption [25].

ST-LSTM: In 2019, [26] used the Spatiotemporal-Long Short Term Memory network (ST-LSTM) for motion planning. The steering angle control model for autonomous vehicles utilized the ST-LSTM architecture, which consists of four Conv-LSTMs [24]. These Conv-LSTMs are responsible for processing the multi-frame feature information. Additionally, a 3D convolutional layer is applied to extract spatiotemporal information from the processed features. The final step involves passing the extracted information through two fully connected layers.

BO-ST-LSTM: recently, [27] used Bayesian Optimization in their network. Discovering the most effective deep neural network architecture and fine-tuning hyperparameters can be a resource-intensive endeavor. However, in this particular study, the optimization of the ST-LSTM network was accomplished through the utilization of Bayesian Optimization (BO). The primary objective was to construct a dependable model capable of accurately predicting the steering angle in self-driving cars. By employing Bayesian

Optimization, the researchers aimed to enhance the reliability and robustness of the steering angle prediction model, acknowledging the high costs associated with architecture optimization and hyperparameter tuning. This model performed more accurately on a public dataset than previous classical end-to-end models. However, implementing an optimization process on hyper-parameter requires training and evaluating 25 different ST-LSTM models, consisting of Conv-LSTM. While BO-ST-LSTM demonstrated superior performance compared to previous models, it comes at a higher cost. This is due to the slower processing speed and increased memory requirements of Conv-LSTM compared to other networks. Additionally, the training process for this model necessitates 25 iterations, further contributing to the higher cost associated with its implementation.

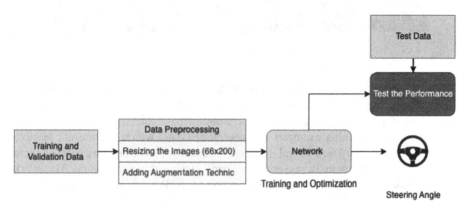

Fig. 3. Overview of Our Work

3   Implementation Details

3.1 Data Preprocessing

Figure 3 shows an overview of our work. The model is trained and assessed using the updated version of Sully Chen Public Dataset [28]. This dataset consists of 63.825 images of the front camera, paired with its steering wheel angles. Figure 4 demonstrate 4 samples from this dataset. To ensure fair comparison with other models and due to computational constraints, 62% of the dataset (39,713 samples) was utilized, with 64% allocated for training data, 16% for validation, and 20% for testing. The original image's dimensions are 455×256 pixels, and the image size is reduced to 200×66 pixels.

3.2 Augmentation

We employed augmentation techniques for two main reasons. Firstly, the dataset suffered from a significant imbalance, where the majority of the steering wheel data corresponded to a neutral position (i.e., 0). This was primarily due to the prevalence of straight sections on highway routes, with a comparatively small portion representing curving streets.

Fig. 4. Four samples from Sully Chen Public Dataset

This means that if we use this dataset, our model will be biased toward driving straight. Augmentation techniques are used to increase curved road data. Secondly, a typical convolutional neural network can have up to a million parameters, and fine-tuning them needs millions of training examples of uncorrelated data, which is often not feasible and can be expensive. Deep neural networks frequently overfit the data because there is a lack of data. One way to prevent over-fitting is through augmentation. This work uses translation, brightness, darkness, and flipping methods for augmentation techniques [29].

3.3 Network Architecture

Figure 5 show the architecture of our work. This model of ours is inspired by NVIDIA's CNN architecture [20]. This network consists of two main parts: convolution layers and fully connected layers. The primary purpose of convolutional layers is to identify and extract critical features from the input data. On the other hand, the role of fully connected layers predominantly revolves around using the extracted features to predict the steering angle. However, they are not clearly separated from one another because this model follows an end-to-end approach. In addition, a series of Batch Normalization layers (BN) have been used to normalize each layer's inputs. The network becomes faster and more stable by normalizing input layers with new centrality and new scaling [30]. In order to avoid data overfitting in fully connected layers, the dropout [10] layer is used, which has a neuron removal rate of 0.2. Equivalently, DropBlock [31] is used to reduce overfitting in convolution layers. While dropout is commonly employed as a regularization technique for fully connected layers, its effectiveness is often reduced when applied to convolutional layers. This layer is similar to dropout, except that the contiguous units are dropped together from the feature map. In this network, this layer removes blocks with a probability of 25%.

More precisely, five convolution layers are used. The first two convolution layers include 24 and 36 5x5 filters with a stride of 2, respectively. The output from each layer is processed through a batch normalization layer, which serves to normalize the data distribution before passing it to the subsequent layer. Then there is a convolution layer with 48 5x5 filters and a stride of 2; the output of this layer passes through the DropBlock layer before entering batch normalization. Next, there are two other convolution layers, each of which has 64 3x3 filters with a stride of 1. The first output passes through the batch normalization layer, and the second one passes through the DropBlock layer and the batch normalization layer, respectively. The final output of this part is flatted before entering the fully connected layers. These layers contain 1164, 200, 50, 10, and 1 neurons, respectively. The outputs from the first three fully connected layers undergo subsequent processing through the dropout and batch normalization layers before being fed into the next fully connected layer. Similarly, the output from the fully connected layer consisting of ten neurons is passed through the batch normalization layer and serves as input for the final layer, which generates the output or steering angle. Throughout these layers, except for the last one, the activation function employed is ReLU.

Dropout and DropBlock techniques in the network architecture, together with the augmentation technique used in data preprocessing, reduce the problem of overfitting. Also, by observing the output of the convolution layers, it can be concluded that the network of this model has the ability to determine the road and the path in which the car can move, as well as other vehicles and road lanes if present. In general, the basis of this network is to determine the factors that affect the steering angle. Table 1 summarizes the architectures for 5 networks.

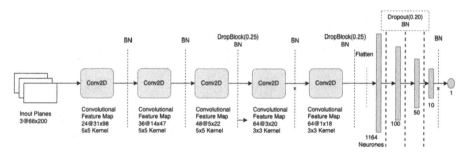

Fig. 5. Architecture of Network

3.4 Loss Function and Optimization

This research aims to train the model to produce the right steering angle. Due to the nature of the problem, we need to compare the result produced by the model with the actual value of the steering angle.

As a result, Mean Square Error (MSE) and Mean Absolute Error (MAE) are used as the loss functions. Its two equations are shown below.

$$\text{MSE} = \frac{1}{n} \sum (y_i - \hat{y}_1)^2 \tag{1}$$

$$\text{MAE} = \frac{1}{n} \sum |y_i - \hat{y}_i| \tag{2}$$

Table 1. Architectures for 5 networks

PilotNet	J-Net	St-LSTM	BO-ST-LSTM	Ours
2017	2019	2019	2021	2022
Normalization	Normalization	Normalization	Bayesian optimization	Conv2D-layer
Conv2D-layer	Conv2D-layer	Conv-LSTM	Normalization	BN
Conv2D-layer	Max-Pooling2D	BN	Conv-LSTM	Conv2D-layer
Conv2D-layer	Conv2D-layer	Conv-LSTM	BN	BN
Conv2D-layer	Max-Pooling2D	BN	Conv-LSTM	Conv2D-layer
Conv2D-layer	Conv2D-layer	Conv-LSTM	BN	DropBlock
Flatten	Max-Pooling2D	BN	Conv-LSTM	BN
Dense	Flatten	Conv-LSTM	BN	Conv2D-layer
Dense	Dense	BN	Conv-LSTM	BN
Dense		Conv3D	BN	Conv2D-layer
		Max-Pooling3D	Conv3D	DropBlock
		Flatten	Max-Pooling3D	BN
		Dense	Flatten	Flatten
			Dense	Dense(1164)
			Dropout(0.28)	Dropout(0.20)
				BN
				Dense(200)
				Dropout(0.20)
				BN
				Dense(50)
				Dropout(0.20)
				BN
				Dense(10)
				BN
Output-Layer	Output-Layer	Output-Layer	Output-Layer	Output Layer

For loss optimization, we used "ADAM" as an optimizer for our Deep Convent model with a learning rate of 0.001. Other values are the defaults of the Keras ADAM, which showed promising results during training.

4   Results and Discussion

In the study conducted by [27], four distinct architectures were implemented and trained for 15 epochs. The training process utilized batch sizes of 50 and followed specific configurations based on their respective definitions. The architectures employed were

PilotNet [20], J-net [21], a modified version of the ST-LSTM proposed by [26], and Bo-ST-LSTM [27]. We used the same properties and trained our model.

The results of the network on both training and validation data are presented in Table 2. Among the five networks examined, our network exhibits the most favorable performance on the validation data. Conversely, the other four networks display symptoms of overfitting, as they demonstrate minimal error on the training data but substantially higher error on the validation data. In contrast, our network demonstrates the highest error on the training data and the lowest error on the validation data compared to the other networks. The reason for the higher MSE/MAE in the training phase is attributed to the increased complexity of the training dataset, despite having the same size of training data as other networks. Other models simply copied curved data directly without modifying the input image, which may have helped them address the imbalanced dataset. In contrast, our approach prioritized the creation of a more intricate and diverse training dataset by incorporating augmentation techniques. This augmentation technique likely introduced variations and transformations to the original data, resulting in a more complex training set. Although this increased complexity may have contributed to higher errors during training, it can also provide the potential for improved generalization and performance on unseen data.

Table 2. The table presents the average prediction performance of five different architectures on both the training and validation data. The most favorable values are highlighted in bold.

		PilotNet	J-Net	ST-LSTM	BO-ST-LSTM	Ours
		(2017)	(2019)	(2019)	(2021)	(2022)
Training	MSE	0.0209	**0.0114**	0.0405	0.1831	0.4972
	MAE	0.0870	**0.0679**	0.1181	0.3798	0.4154
Validation	MSE	0.6814	0.5842	0.6139	0.5019	**0.1053**
	MAE	0.4409	0.4262	0.4710	0.4042	**0.1482**

Then we tested the network with test data (Table 3). Although J-Net produced the slightest error on the training data, it has the highest error on the test data. Our network has the least error on the test data, producing the highest error on the training data. The network's performance has been improved by about 60%, and the problem of overfitting has also been reduced.

4.1 Ablation Studies

To analyze the influence of different network layers, we conducted an experiment to examine the effects of removing each layer individually. The layers subjected to removal were DropBlock, Dropout, and Batch Normalization. Figure 6 shows how much the network's result will change if we remove each layer. Our findings indicate that the removal of these layers had distinct impacts on the network's performance. Specifically, removing the DropBlock layer resulted in a 10% decrease in performance. Removing the

Table 3. The Table shows the five network's MSE values on training and test data.

	Training Data	**Test Data**
(2017) PilotNet	0.02090	0.28100
(2019) J-Net	**0.01140**	0.32040
(2019) ST-LSTM	0.04050	0.27580
(2021) BO-ST-LSTM	0.18310	0.27000
(2022) Ours	0.491720	**0.10620**

Dropout layer had a more significant effect, leading to a 55% decrease in performance. However, the most substantial impact was observed when the Batch Normalization layer was removed, resulting in a 60% decrease in performance compared to the original architecture.

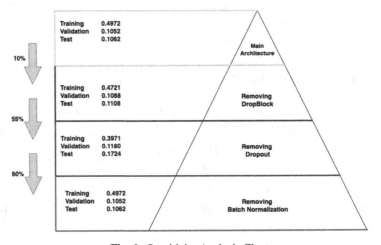

Fig. 6. Sensitivity Analysis Chart

5 Conclusion

This paper aims to predict steering angle using just photos of the road taken by a camera placed on the car's dashboard. We introduce a comprehensive approach for generating control commands in an end-to-end manner. The convolutional neural network takes in the raw frames of the road image as input and produces the steering wheel angle as output. The training of this network was conducted using the publicly available Sully Chen dataset. A challenge for this approach is how to encode or extract the influencing features on control commands to convert them into appropriate control commands such as steering angle. Therefore, network architecture and data preprocessing are critical in this architecture. In this work, we designed a novel network that produces promising results compared to previous networks. In addition, the previous end-to-end models suffer from the overfitting problem; this problem has been largely resolved with the help of the new network architecture as well as different preprocessing of the data in this work.

References

1. Krizhevsky, A., Sutskever, I., Hinton, G.E.: ImageNet classification with deep convolutional neural networks. Adv. Neural. Inf. Process. Syst. **25**, 1–9 (2012)
2. Andrychowicz, M., et al.: Learning Dexterous In-Hand Manipulation (2018)
3. Goldberg, Y.: Neural Network Methods for Natural Language Processing, Morgan & Claypool Publishers, Ed., p. 309 (2017)
4. O'Shea, K., Nash, R.: An Introduction to Convolutional Neural Networks, ArXiv e-prints (2015)
5. Sherstinsky, A.: Fundamentals of recurrent neural network (RNN) and long short-term memory (LSTM) network. Physica D **404**, 132306 (2018)
6. Goodfellow, I.J., et al. : Generative Adversarial Networks. In: Advances in Neural Information Processing Systems, vol. 27, Z. Ghahramani, M. Welling, C. Cortes, N. Lawrence and K. Weinberger, Eds., Curran Associates, Inc. (2014)
7. Zhang, Q., Pan, W., Reppa, V.: Model-reference reinforcement learning for collision-free tracking control of autonomous surface vehicles. IEEE Trans. Intell. Transp. Syst. **23**(7), 1558–1616 (2020)
8. Grigorescu, S., Trasnea, B., Cocias, T., Macesanu, G.: A survey of deep learning techniques for autonomous driving. J. Field Robot. **37**(3), 362–386 (2020)
9. Ni, J., Shen, K., Chen, Y., Cao, W., Yang, S.X.: An Improved deep network-based scene classification method for self-driving cars. IEEE Trans. Instrum. Meas. **71**, 1–14 (2022)
10. Srivastava, N., Hinton, G., Krizhevsky, A., Sutskever, I., Salakhutdinov, R.: Dropout: a simple way to prevent neural networks from overfitting. J. Mach. Learn. Res. **15**(56), 1929–1958 (2014)
11. Zhao, J., Xie, B., Huang, X.: Real-time lane departure and front collision warning system on an FPGA. In: IEEE High Performance Extreme Computing Conference (HPEC), pp. 1–5 (2014)
12. Li, C., Wang, J., Wang, X., Zhang, Y.: A model based path planning algorithm for self-driving cars in dynamic environment. In: 2015 Chinese Automation Congress (CAC), pp. 1123–1128, (2015)
13. Wang, S., Lin, F., Wang, T., Zhao, Y., Zang, L., Deng, Y.: Autonomous Vehicle Path Planning Based on Driver Characteristics Identification and Improved Artificial Potential Field, Actuators, vol. 11 (2022)
14. Kong, J., Pfeiffer, M., Schildbach, G., Borrelli, F.: Kinematic and dynamic vehicle models for autonomous driving control design. In: 2015 IEEE Intelligent Vehicles Symposium (IV), pp. 1094–1099 (2015)
15. Wang, D., Feng, Q.: Trajectory planning for a four-wheel-steering vehicle. In: Proceedings 2001 ICRA. IEEE International Conference on Robotics and Automation (Cat. No.01CH37164), vol. 4, pp. 3320–3325 vol.4 (2001)
16. Chen, J., Li, S.E., Tomizuka, M.: Interpretable end-to-end urban autonomous driving with latent deep reinforcement learning. IEEE Trans. Intell. Transp. Syst. **23**(6), 5068–5078 (2022)
17. Chen, S., Wang, M., Song, W., Yang, Y., Li, Y., Fu, M.: Stabilization approaches for reinforcement learning-based end-to-end autonomous driving. IEEE Trans. Veh. Technol. **69**(5), 4740–4750, (2020)

18. Chen, Z., Huang, X.: End-to-end learning for lane keeping of self-driving cars. In: 2017 IEEE Intelligent Vehicles Symposium (IV), pp. 1856–1860 (2017)

19. Bicer, Y., Alizadeh, A., Ure, N.K., Erdogan, A., Kizilirmak, O.: Sample efficient interactive end-to-end deep learning for self-driving cars with selective multi-class safe dataset aggregation. In: 2019 IEEE/RSJ International Conference on Intelligent Robots and Systems (IROS), pp. 2629–2634 (2019)

20. Bojarski, M., et al.: Explaining How a Deep Neural Network Trained with End-to-End Learning Steers a Car (2017)

21. Kocić, J., Jovičić, N., Drndarević, V.: An end-to-end deep neural network for autonomous driving designed for embedded automotive platforms, sensors, (2019)

22. Xin, M., Wang, Y.: Research on image classification model based on deep convolution neural network. EURASIP J. Image Video Process. **2019**(1), 40 (2019)

23. Lindemann, B., Müller, T., Vietz, H., Jazdi, N., Weyrich, M.: A survey on long short-term memory networks for time series prediction. Procedia CIRP **99**, 650–655 (2020)

24. Shi, X., Chen, Z., Wang, H., Yeung, D.Y., Wong, W.K., Woo, W.C.: Convolutional LSTM network: a machine learning approach for precipitation nowcasting. In: NIPS'15: Proceedings of the 28th International Conference on Neural Information Processing Systems, vol. 1, pp. 802–810 (2015)

25. Pfeuffer, A., Dietmayer, K.: Separable convolutional LSTMs for faster video segmentation. In: 2019 IEEE Intelligent Transportation Systems Conference (ITSC), pp. 1072–1078 (2019)

26. Bai, Z., Cai, B., ShangGuan, W., Chai, L.: Deep learning based motion planning for autonomous vehicle using spatiotemporal LSTM network, pp. 1610–1614, (2019)

27. Riboni, A., Ghioldi, N., Candelieri, A., Borrotti, M.: Bayesian optimization and deep learning forsteering wheel angle prediction (2021)

28. Chen, S.: driving-datasets (2018). (https://github.com/SullyChen/driving-datasets)

29. Shorten, C., Khoshgoftaar, T.M.: A survey on image data augmentation for deep learning. J. Big Data **6**(1), 60 (2019)

30. Ioffe, S., and Szegedy, C.: Batch normalization: accelerating deep network training by reducing internal covariate shift. In: Proceedings of the 32nd International Conference on Machine Learning, vol. 37, pp. 448--456 (2015)

31. Ghiasi, G., Lin, T.-Y., Le, Q.V.: Dropblock: A regularization method for convolutional networks, Advances in neural information processing systems, vol. 31 (2018)

Fractal-Based Spatiotemporal Predictive Model for Car Crash Risk Assessment

Behzad Zakeri[1(✉)] and Pouya Adineh[2]

[1] School of Mechanical Engineering, College of Engineering, University of Tehran, Tehran, Iran
Behzad.Zakeri@ut.ac.ir
[2] Department of Computer Engineering, Faculty of Engineering, Islamic Azad University South Tehran Branch, Tehran, Iran

Abstract. Car collisions are a noteworthy public safety concern, with a significant number of fatalities and injuries worldwide each year. In this study, we developed a spatiotemporal prediction model for car collision risk by combining fractal theory and deep learning techniques. We collected and analyzed car collision data for the United States between 2016 and 2021, and used fractal theory to extract effective parameters and the Hurst exponent to analyze temporal patterns. Finally, we trained a predictive model using Generative Adversarial Networks (GANs) and Long Short-Term Memory (LSTM) networks. Our results demonstrate the potential of fractal theory and deep learning techniques for developing accurate and effective spatiotemporal prediction models, which can be utilized to identify areas and time periods of heightened risk and inform targeted intervention and prevention efforts.

Keywords: Car collision · Spatiotemporal prediction · Fractal theory · Deep learning

1 Introduction

Car collisions is a signficant public safety concern, with an estimated 1.35 million fatalities worldwide each year, according to the World Health Organization recordes [12]. In the United States alone, there were more than 38,000 fatalities and 4.4 million injuries due to motor vehicle crashes in 2019 [1]. Despite advances in automotive technology and safety regulations, car collisions continue to occur at an alarming rate, and identifying ways to predict and prevent them remains a critical research goal.

To this end, spatiotemporal prediction models have emerged as a promising tool for identifying areas and time periods of heightened risk. Such models rely on the collection and analysis of large amounts of data to identify patterns and correlations that can be used to predict future collisions.

In recent years, there has been growing interest in the use of fractal theory and related mathematical techniques for analyzing car collision data. Fractal theory involves the study of complex systems that exhibit self-similarity at different scales, and has been used to analyze a wide range of physical and natural phenomena. In the context of car collisions, fractal theory has been used to identify correlations between collision

M. Ghatee and S. M. Hashemi (Eds.): ICAISV 2023, CCIS 1883, pp. 205–215, 2023.
https://doi.org/10.1007/978-3-031-43763-2_13

frequency and various environmental factors, such as road geometry, traffic flow, and weather conditions [15].

In addition to fractal theory, the Hurst exponent has also been used to analyze car collision data. The Hurst exponent is a measure of long-term memory in a time series, and has been used to identify patterns and trends in various complex systems. In the context of car collisions, the Hurst exponent has been used to identify correlations between collision frequency and various temporal factors, such as time of day, day of the week, and season [3].

In addition to our use of fractal theory and the Hurst exponent, we also employed the singular value decomposition (SVD) method to extract spatiotemporal features from our car collision dataset. SVD is a powerful technique for dimensionality reduction that can help to better extract spatiotemporal features and improve control on effective parameters. By using SVD, we were able to identify the most important features in our dataset, which enabled us to develop more accurate spatiotemporal prediction models.

Despite these advances, developing accurate and effective spatiotemporal prediction models remains a challenging task. One promising approach involves the use of deep learning techniques, such as Generative Adversarial Networks (GANs) and Long Short-Term Memory (LSTM) networks. GANs have been used to generate synthetic collision data, which can be used to augment the available training data and improve the accuracy of predictive models [8]. LSTM networks, on the other hand, have been used to analyze temporal patterns and identify trends in car collision data [11, 16].

In this study, we collected and analyzed car collision data for the United States between the years 2016 to 2021, which includes more than 10 million collisions [1]. We used fractal theory, the Hurst exponent, and SVD to extract effective parameters and spatiotemporal features, and found that the dimension of the physics is an important factor for predicting collision risk. Finally, we combined GAN and LSTM networks to train a spatiotemporal prediction model for car collision risk, which can be used to identify areas and time periods of heightened risk and inform targeted intervention and prevention efforts.

Overall, our study contributes to the growing body of research on car collision prediction and prevention, and demonstrates the potential of combining fractal theory, the Hurst exponent, and SVD with deep learning techniques for developing accurate and effective spatiotemporal prediction models.

2   Methodology

2.1   Fractal Theory

The fractal theory was introduced by Benoit mandelbrot in 1967 [10] to enhance the performance of analyzing complex systems. In fractal theory, the estimation of topological roughness can be calculated numerically by introducing the "Fractal Dimension" (FD). The FD can be considered as a metric to measure the roughness; shapes with higher FD are rougher and vice versa.

To have a better understanding of FD, Koch set gives useful insight. The Koch set is generated by dividing a line segment with length L into 3 distinct segments (L/3 in length), then the middle segment removes and replaced by an equilateral triangle with a side length equal to L/3. Repeating this process gives a self-similar fractal which is shown in Fig. 1.

Dimensionality analysis of the obtained shapes leads to a noninteger dimension which can be calculated by Eq. 1, [14].

$$d = \frac{\ln m}{\ln r} \tag{1}$$

where d, m and r are the similarity dimension, number of copies and scale factor, respectively.

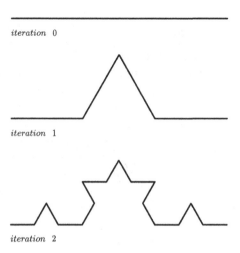

iteration 0

iteration 1

iteration 2

Fig. 1. The Koch set

It is prominent that in each iteration, a segment line is divided by 3. The final result in each iteration is four segmented lines corresponding to the primer segmented line. It means m and r, in this case, are 4 and 3, respectively.

Now let us turn our attention to the more practical methods for the characterization of various sets. The correlation dimension D_C, is one of the robust methods for calculation the dimensionality of data sets. This method consists of assuming a hypersphere about an arbitrary data point (i) and letting the radius of the hypersphere (r) grow till all the data points are enclosed. Since in practical cases, the number of data points (n) is finite, many hyperspheres with different initial points are used, and ensemble averaging is implemented. This is the definition of the correlation function, $N(r)$, shown by Eq.2 :

208   B. Zakeri and P. Adineh

$$N(r) = \lim_{n \to \infty} \frac{1}{n^2} \sum_{i \neq j}^{n} \sum_{j}^{n} H(r - |x_i - x_j|) \qquad (2)$$

Where the $H(\xi)$ is the Heaviside operator and is defined as follows:

$$H(\xi) = \begin{cases} 1.0, & \xi \geq 0 \\ 0.0, & \xi < 0 \end{cases} \qquad (3)$$

From Eq.2, it is obtainable that if the limit $r \to 0$, the Eq. 2 is reformed as below:

$$N(r) = C_1 r^{D_C} \qquad (4)$$

So, it is concluded that,

$$\log N = D_C \log r + C_2 \qquad (5)$$

Looking more precisely into Eq. 5, it is obvious that if we plot $N(r)$ versus r in the log-log plot, the slope of the curve is the correlation dimension, D_C, where demonstrated in Fig. 2.

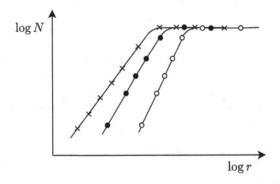

Fig. 2. Correlation Dimension D_C

The functionality of this method to compute the D_C is flawlessly smooth for numerically evaluated functions; However, when it comes to real-world datasets, it is not as straightforward as before.

One of the challenging issues in tackling real-world data sets is the unknown number of state variables for characterization. However, it has been shown that time-delayed reading of only one temporal measurement and construction of pseudo-phase-space leads to the reconstruction of the topology of the system attractor [13].

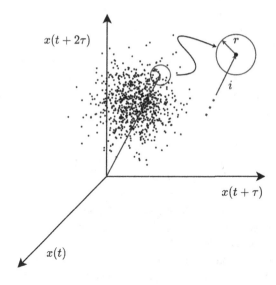

Fig. 3. Phase-space representation of signal $X(t)$

Assuming in the under-study system a feature $\chi(t)$ has been measured, it is straightforward to read $\chi(t)$, $\chi(t + \tau)$, $\chi(t + 2\tau)$, ..., where the τ is the arbitrary time delay in the intended system.

For the determination of the D_C, an arbitrary dimension is chosen, for instance, $3D$ which is shown schematically in Fig. 3. As described before, a sphere with radius r is considered about the point-i and Eq. 1 is applied. Then, the $\log - \log$ graph is drawn similar to Fig. 2, and the slope of graph $D(3)$ is calculated. This process is continued till by the increment in the arbitrary dimension, the slope of the $\log - \log$ graph remains constant. The last dimension, which makes the $\log - \log$ graph constant is the real dimension of the under-study system.

It is worth emphasizing that in the cases in which data sets are random processes, there is no finite dimension, and the slope of the $\log - \log$ graph never converges to a specific number. To study the system it is inevitable to know that the system, is a deterministic, random process, or chaotic. For this purpose, the Hurst dimension is used which is described below.

2.2 Hurst Exponent

One of the significant fractal dimensions called the Hurst dimension (D_H) was introduced by the work of Fedor in 1988 [6]. Hurst demonstrated that lots of time-varying chaotic processes $\xi(t)$, of record length τ, can be correlated by :

$$\frac{R(\tau)}{S(\tau)} = (\tau/2)^{D_H} \tag{6}$$

where,

$$R(\tau) \overset{\Delta}{=} \max_{t \in \tau} \chi(t; \tau) - \min_{t \in \tau} \chi(t; \tau) \tag{7}$$

$$\chi(t;\tau) = \int_{t-\tau}^{t} [\xi(t') - \bar{\xi}(\tau)]dt' \tag{8}$$

$$\bar{\xi} = \frac{1}{\tau} \int_{t-\tau}^{t} \xi(t')dt' \tag{9}$$

and,

$$S = \left\{ \frac{1}{\tau} \int_{t-\tau}^{t} [\xi(t') - \bar{\xi}]^2 dt' \right\}^{\frac{1}{2}} \tag{10}$$

In Eqs. 6–10, $\chi(t;\tau)$ is the collected temporal variation of $\xi(t)$ about its mean, $\bar{\xi}$; $R(\tau)$ is the difference between the maximum and minimum value of $\chi(t;\tau)$ in interval τ, and S is the standard deviation of $\xi(t)$.

The Hurst dimension quantity is bounded between 0 and 1 ($0 < D_H \leq 1$), where each interval demonstrates a specific property in the physical system. In the case $D_H = 0.5$, the system is a random process, while $D_H > 0.5$ means there is some underlying structure in the system. A fully deterministic system has $D_H = 1$ and most natural processes have $D_H \simeq 0.72$.

2.3 Singular Value Decomposition

Singular Value Decomposition (SVD) is a specific type of matrix decomposition that is available for any complex-valued matrix. The SVD of an arbitrary matrix X of size $m \times n$, contains three matrices (U, Σ, and V). Where U and V are orthogonal matrices with size $m \times m$, while Σ is a diagonal matrix filled by singular values [2].

$$X = U\Sigma V^* \tag{11}$$

where $U \in \mathbb{C}^{n \times n}$ and $V \in \mathbb{C}^{m \times m}$ are unitary matrices with orthonormal columns, and $\Sigma \in \mathbb{R}^{n \times m}$ is a matrix with real, nonnegative entries on the diagonal and zeros off the diagonal. Here $*$ denotes the complex conjugate transpose.

After the definition of the SVD, to take advantage of the efficient property of this matrix decomposition method it is worth emphasizing to define a theorem as below :

Theorem 1 (Eckart-Young [4]). *The optimal rank-r approximation to X, in a least-squares sense is given by the rank-r SVD truncation* $\tilde{X}$:

$$\underset{\tilde{X}, s.t. rank(\tilde{X})=r}{\arg\min} \ \ ||X - \tilde{X}||_F = \tilde{U}\tilde{\Sigma}\tilde{V}^* \tag{12}$$

Here, $\tilde{U}$ and $\tilde{V}$ denote the first r leading columns of U and V, and $\tilde{\Sigma}$ contains the leading $r \times r$ sub-block of Σ. Also $||.||_F$ is the Frobenius norm.

Using theorem 1, it is straightforward to truncate the matrix X, when truncated matrices $\tilde{U}$, $\tilde{\Sigma}$, and $\tilde{V}$ are available. For truncation values r that are smaller than the number of nonzero singular values (i.e., the rank of X), the truncated SVD only approximates X :

$$X \approx \tilde{U}\tilde{\Sigma}\tilde{V}^* \tag{13}$$

2.4 Deep Learning

In this section, with the combination of two networks, Generative Adversarial Network (GAN) and Long-Short Term Memory (LSTM), the recurrent-generative adversarial network (RGN) propose. The idea of presenting the RGN is developing an intelligent network which can learn the spatiotemporal property of collision risk regarding to recognising different condition to get the maximum performance. With benefiting of the LSTM and GAN, the RGN got tuned to predict the most susceptible time and location for probable collisions. Rest of this part is explaining the combination of LSTM and GAN networks, and the way that this arcitecture works.

The RGN is performed by a specific kind of GAN called conditional GAN (CGAN) to generate target results by considering particular conditions. This CGAN in its generator and discriminator parts utilized the RNN structure, which in this study is LSTM. Figure 4 depicts the generator and discriminator combined with LSTM. The recurrent conditional GAN (RCGAN) is in charge to learn the pattern of the traffic dynamics regarding the risk of collision and generate the optimum spatiotemporal results. To get an accurate generation by the generator, the discriminator has been pre-trained by several datasets extracted from the fractal theory analysis cleaned by Singular Value Decomposition (SVD) method.

The discriminator is trained to minimize the negative cross-entropy among generated traffic collision of each time-steps and the corresponding actual collision from the reference dataset. With considering the $RNN(X)$ the vector regarding T output from the RNN giving a sequence of T vectors $\{x_t\}_{t=1}^{T}$ ($x_t \in \mathbb{R}^d$), and the $CE(\alpha, \beta)$ the average-cross-entropy between sequences α and β. In this way, the loss for a pair of

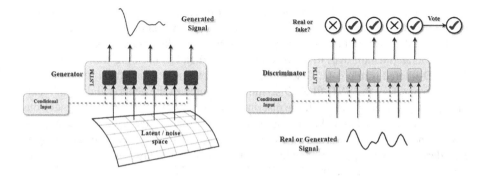

Fig. 4. Combination of LSTM and GAN

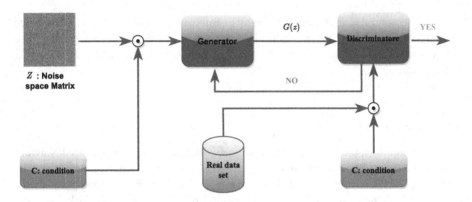

Fig. 5. Schematic diagram of the GAN architecture

$\{X_n, y_n\}$ (with $X_n \in \mathbb{R}^{T \times d}$ and $y_n \in \{1, 0\}^T$) is defined in Eq. 14 as follow :

$$D_{loss}(X_n, y_n) = -CE(RNN_D(X_n; y_n)) \tag{14}$$

In real-valued signals, y_n is a vector containing 1 s or 0 s for generated signals. In each minibatch, the discriminator considers both real and generated signals. The aim of the generator is to trick the discriminator into considering the generated signals as real-valued signals, and this purpose is acquired by minimizing the average of the negative cross-entropy among the discriminator's judgment on generated signals and real signals (known as 1 s) shown in the Eq. 15:

$$G_{loss}(Z_n) = D_{loss}(RNN_G(Z_n), 1) = -CE(RNN_D(RNN_G(Z_n)), 1) \tag{15}$$

In the Eq. 15, Z_n is a signal containing T independent sample points from noise space Z, and since $Z = \mathbb{R}^m$ then $Z_n \in \mathbb{R}^{T \times m}$.
In this study, the condition c_n contains the SVD filter which only passes the main spatiotemporal feature to minimize the computational costs.

In the conventional GAN, the generator unit takes the input from the latent/noise space, the RCGAN, however, gets an input conducted by a combination of noise and conditions. The generator tries to generate the most similar results to the real-valued data. Generated data moves to the discriminator, which has been pre-trained by previous SVD-filtered traffic data. Discriminator takes two inputs, one from the generator, and the other from the SVD filter. Discriminator compares the generated data and real-valued data from the database. If the discriminator detects that the imported signal from the generator is artificial and does not match with features of the database, returns the error to the generator to tune the weights and improve its performance, else the signal considers the correct prediction. Meaning the generator is fully trained, and ready to predict the spatiotemporal traffic collision properties. The whole structure of the discussed architecture is shown in the Fig. 5.

3   Results

In this section, the recurrent-generative predictive model performance has been evaluated. The accuracy of the model has been determined by utilizing the maximum mean discrepancy (MMD) metric. Also, the convergence of the learning procedure is demonstrated by using the MMD^2 for the Epoch number.

3.1   Maximum Mean Discrepancy

It is necessary to evaluate the learning procedure of the GAN network. To this end, the GAN network considers successful in learning if it could learn the distribution of patterns in the real-valued data. For this purpose, generated data confirmed by the discriminator has been studied. In these kinds of problems, it is conventional to use maximum mean discrepancy (MMD) [5,7]. The MMD metric has been used as a training objective for moment matching networks [9]. MMD is in charge of asking if the sample belongs to the real-value database or is the generated sample. In this study, the squared of MMD (MMD^2) has been used, and instead of the inner product between functions of samples (real-valued and generated), a kernel has been utilised. Regarding a kernel $K : X \times Y \to \mathbb{R}$, and samples $\{x_i\}_{i=1}^{N}$ and $\{x_j\}_{j=1}^{M}$, the estimated value of MMD^2 is defined as follow:

$$
\widehat{MMD}_u^{\,2} = \frac{1}{n(n-1)} \sum_{i=1}^{n} \sum_{j \neq i}^{n} K(x_i, x_j) - \frac{2}{mn} \sum_{i=1}^{n} \sum_{j=1}^{m} K(x_i, y_j)
$$
$$
+ \frac{1}{m(m-1)} \sum_{i=1}^{m} \sum_{j \neq i}^{m} K(y_i, y_j) \tag{16}
$$

It is still a challenging wide field of study to chose proper kernels among time series. The alignment of time series can be named as the technique that is using to tackle this challenge. In this work, also, by fixing the time axis, real-valued data and sensors feedback aligned. The time series reformed as vectors for comparisons, and the radial basis function (RBF) kernel with taking advantage of ℓ_2-norm among vectors as $K(x,y) = \exp(-\frac{\|x-y\|^2}{(2\sigma)^2})$. To find the optimal bandwidth (σ), the estimator of the t-statistic of the power of the MMD is maximised. Where $\hat{t} = \frac{\widehat{MMD}^2}{\sqrt{Var}}$, the Var is the asymptotic variance of the estimator of MMD^2.

Figure 6 demonstrates the MMD^2 convergence and Generator/Discriminator losses. As it is shown in Fig. 6a the convergence in MMD^2 has acquired in epoch 50, and after this epoch number, its value fluctuates around 0.001. This number depicts the acceptable accuracy of the predictive model in the prediction of spatiotemporal patterns. Figure 6b shows the convergence of generator and discriminator losses. It is obvious that both losses converged acceptably, although both have considerable fluctuation and it is much more in discriminator loss. The reason for this fluctuation is generator could successfully trick the discriminator. It is worth emphasizing, although both G and D have fluctuation, this fluctuation is around near-zero value.

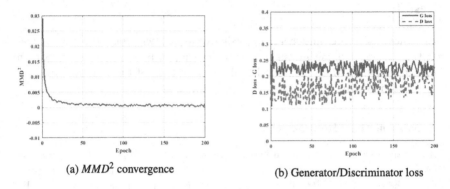

(a) MMD^2 convergence

(b) Generator/Discriminator loss

Fig. 6. Diagrams of metrics for the learning procedure

4   Conclusion

In this paper, the ability of the combination of fractal theory and the recurrent-generative predictive network has been shown for car collision risk assessment. Also, to improve the power of the network data SVD technique has been utilized. The MMD metric demonstrates the robust ability of the proposed method. It is worth emphasizing that there are several other methods that could be considered to improve the conducted result which could be done in further work.

References

1. Adeyemi, O.J., Paul, R., Arif, A.: An assessment of the rural-urban differences in the crash response time and county-level crash fatalities in the united states. J. Rural Health **38**(4), 999–1010 (2022). https://doi.org/10.1111/jrh.12627
2. Brunton, S.L., Kutz, J.N.: Data-Driven Science and Engineering: Machine Learning, Dynamical Systems, and Control. Cambridge University Press, 2 edn. (2022). https://doi.org/10.1017/9781009089517
3. Chang, F., Huang, H., Chan, A.H., Shing Man, S., Gong, Y., Zhou, H.: Capturing long-memory properties in road fatality rate series by an autoregressive fractionally integrated moving average model with generalized autoregressive conditional heteroscedasticity: A case study of florida, the united states, 1975–2018. J. Safety Res. **81**, 216–224 (2022). https://doi.org/10.1016/j.jsr.2022.02.013
4. Eckart, C., Young, G.: The approximation of one matrix by another of lower rank. Psychometrika **1**(3), 211–218 (1936). https://doi.org/10.1007/BF02288367
5. Esteban, C., Hyland, S.L., Rätsch, G.: Real-valued (medical) time series generation with recurrent conditional gans (2017). https://arxiv.org/abs/1706.02633
6. Feder, J.: Fractals. Springer, New York, NY (November 2013). https://doi.org/10.1007/978-1-4899-2124-6
7. Huang, J., Gretton, A., Borgwardt, K., Schölkopf, B., Smola, A.: Correcting sample selection bias by unlabeled data. In: Schölkopf, B., Platt, J., Hoffman, T. (eds.) Advances in Neural Information Processing Systems. vol. 19. MIT Press (2006). https://proceedings.neurips.cc/paper/2006/file/a2186aa7c086b46ad4e8bf81e2a3a19b-Paper.pdf

8. Khoshrou, M.I., Zarafshan, P., Dehghani, M., Chegini, G., Arabhosseini, A., Zakeri, B.: Deep learning prediction of chlorophyll content in tomato leaves. In: 2021 9th RSI International Conference on Robotics and Mechatronics (ICRoM), pp. 580–585 (2021). https://doi.org/10.1109/ICRoM54204.2021.9663468

9. Li, Y., Swersky, K., Zemel, R.: Generative moment matching networks. In: Bach, F., Blei, D. (eds.) Proceedings of the 32nd International Conference on Machine Learning. Proceedings of Machine Learning Research, vol. 37, pp. 1718–1727. PMLR, Lille, France (2015). https://proceedings.mlr.press/v37/li15.html

10. Mandelbrot, B.: How long is the coast of britain? statistical self-similarity and fractional dimension. Science **156**(3775), 636–638 (1967). https://doi.org/10.1126/science.156.3775.636, https://www.science.org/doi/abs/10.1126/science.156.3775.636

11. Marcillo, P., Valdivieso Caraguay, n.L., Hernández-Álvarez, M.: A systematic literature review of learning-based traffic accident prediction models based on heterogeneous sources. Appl. Sci. **12**(9) (2022). https://doi.org/10.3390/app12094529 ,https://www.mdpi.com/2076-3417/12/9/4529

12. Organization, W.H.: Global status report on alcohol and health 2018. World Health Organization (2019). https://www.who.int/publications/i/item/9789241565684

13. Packard, N.H., Crutchfield, J.P., Farmer, J.D., Shaw, R.S.: Geometry from a time series. Phys. Rev. Lett. **45**, 712–716 (1980). https://doi.org/10.1103/PhysRevLett.45.712

14. Strogatz, S.H.: Nonlinear dynamics and chaos: with applications to physics, biology, chemistry, and engineering. CRC Press (2018). https://doi.org/10.1201/9780429492563

15. Yan, J., Liu, J., Tseng, F.M.: An evaluation system based on the self-organizing system framework of smart cities: A case study of smart transportation systems in china. Technol. Forecast. Soc. Chang. **153**, 119371 (2020). https://doi.org/10.1016/j.techfore.2018.07.009

16. Zakeri, B., Khashehchi, M., Samsam, S., Tayebi, A., Rezaei, A.: Solving partial differential equations by a supervised learning technique, applied for the reaction-diffusion equation. SN Appl. Sci. **1**(12), 1–8 (2019). https://doi.org/10.1007/s42452-019-1630-x

Author Index

M. Ghatee and S. M. Hashemi (Eds.): ICAISV 2023, CCIS 1883, p. 217, 2023.
https://doi.org/10.1007/978-3-031-43763-2

Printed in the United States
by Baker & Taylor Publisher Services